REFLECTIONS
From an Aging Book Lover

DONALD E. PONT

REFLECTIONS

From an Aging Book Lover

Reflections From an Aging Book Lover

For information about this title or to order other books
and/or electronic media, contact the publisher:

Donald E. Pont
Agingbooklover44@yahoo.com

ISBN: 978-1-7366930-0-1

Printed in the United States of America

Cover and Interior design: 1106 Design

FRONT COVER:
Pictured is the author's grandfather
Rev. Frank E. Pfoutz (1885–1970) with his books.

Dedicated to Jackie, Molly, and Sarah
"Thank You for the Music." (~ABBA)

Contents

Acknowledgments

$\mathcal{M}$any authors have nourished my lifelong passion for reading. I'm particularly beholden to those whom I have quoted or whose works appear in the bibliography. My sincerest gratitude, however, is to my daughters, Molly and Sarah. Sarah inspired me to examine my years of reading and put pen to paper. Molly patiently corrected my misplaced commas and mended my fractured sentences—I can't imagine having a better editor.

A Good Book

There is no frigate like a book
To take us lands away,
Nor any coursers like a page
Of prancing poetry.
—**Emily Dickinson**

*A*s the saying goes: "There's nothing quite like a good book." After a lifetime of reading, I wholeheartedly agree. A great book has a rich, inner life that is seldom found even in a fine film or play.

You may have heard an enthralled reader claim, "That book changed my life!" While I've never been entirely sure what to make of this assertion, I am certain that reading can help answer mysteries in your own life—it can even change your worldview. When I was younger, I began to question how a small movement around an impoverished Nazarene who was crucified somehow morphed into the

Catholic Church. Only after immersing myself in books about early Christianity and the historical Jesus did I come to understand this seemingly improbable happening. I definitely never looked at the world quite the same again after reading Jared Diamond's *Guns, Germs, and Steel*, and I absolutely never viewed my plate in the same way after reading Michael Pollan's *The Omnivore's Dilemma*. The writings of authors like Lewis Thomas and Edward O. Wilson have helped me contemplate my place in nature— and in the universe. I may have learned as much about compassion from James Herriot, the story-telling Yorkshire veterinarian, as I did from any of my physician mentors.

Thrillers and mysteries are often considered mere diversions; however, exposure to the dark underbelly of society can be tremendously entertaining, particularly when one can experience it from the safety of a favorite chair. But thrillers can be much more than blood, guts, and plot. When Carlos Ruiz Zafón penned *The Shadow of the Wind*, he gave us a mystery cloaked in beautiful prose.

In our daily existence, there's a natural limit to the number of friends, or even acquaintances, from whom we can learn. Works of fiction and nonfiction alike enlarge and enrich this cast of characters in our lives. I can't help but think of certain literary characters as the most interesting people I've never met. Who wouldn't want to consult Sherlock Holmes, have tea with Elizabeth Bennet, or—better yet—be a fly on the wall when Owen Meany was in the room? And it's not only heroic characters like James Fenimore Cooper's Natty Bumppo or Sir Walter Scott's angelic Rebecca who captivate the lucky reader. Scores of us were fascinated with the maddeningly imperfect Olive Kitteridge. Even lesser characters like the Reverend Adolphus Irwine in George Eliot's *Adam Bede* can be so compelling that the story is unimaginable without them. Eliot fills her novels with memorable characters that you come to care about, but the sage, comforting Adolphus may be my favorite. I wish he were my friend. In *The Secret Garden*, Frances Hodgson Burnett must have been inspired when she tenderly crafted Dickon. This gentle creature of the earth possesses a natural wisdom that the story's sheltered and selfish children need to see is even possible.

As wonderful as any fictional character can be, it's often the real life—or larger than life—heroes and villains encountered in nonfiction accounts that can fascinate or influence you the most. The novelist can't create characters any more intriguing than some of those real historical figures about whom gifted authors choose to research and write. Take Joseph Priestly, the iconoclastic minister who was kicked out of England for heresy, and—by the way—discovered oxygen in his spare time! It's not just the modern reader who finds Priestly fascinating. His name appears more times in the famous Jefferson-Adams correspondence than does George Washington's. Thomas Jefferson and John Adams themselves were so accomplished—and led such full lives—that there may never be an end to the books written about them. In his era, Benjamin Franklin—scientist, inventor, statesman, intellectual, diplomat, and writer—was the best-known American in the world. We may never tire of reading about Winston Churchill who made history, then deftly wrote it! Who could make up characters like Eleanor of Aquitaine, Leonardo da Vinci, Theodore Roosevelt, or even George Armstrong Custer?

Speaking of the most interesting people I've never met, that group would also encompass legions of authors. The unassuming cardiologist George Sheehan transformed the running movement into philosophy. In writing so elegantly about our founding fathers, David McCullough and Joseph Ellis taught us what it really means to be American. In *Sapiens*, Yuval Noah Harari brilliantly reminded readers that our history started so much sooner than so many want to admit. Luckily for generations of readers, James Boswell followed Samuel Johnson about London scribbling down nuggets like, "The true measure of a man is how he treats someone who can do him absolutely no good." In *The Lucifer Principle*, Howard K. Bloom devastatingly warned us, "If you can convince enough people of your worldview, no matter how wrong you are, you're right." The eminent historian Will Durant wisely reflected, "We must not expect the world to improve much faster than we do ourselves."

Of course, books really can change the world. Isaac Newton's writings revolutionized thought in the seventeenth century, and in the nineteenth century, the reluctant Charles Darwin turned the world upside down by publishing *On the Origin of Species*. While many in our age treasure reading as a deeply personal experience, we should remember that for centuries, books have been the repositories for civilization's collective wisdom and knowledge. When he sat by the fireplace reading William Shakespeare, the young Abraham Lincoln—who was largely self-educated—was tapping into that collective wisdom.

When the brilliant neurologist and best-selling author Oliver Sacks—whom *The New York Times* called "the poet laureate of medicine"—died, I felt a certain sadness—even loss. His words have enriched my life, and his departure reminded me once again that "there's nothing quite like a good book."

The Missing Link

To discover fossils, he told himself, was to reclaim answers to important questions.
—Anthony Doerr, The Shell Collector

$\mathcal{D}$uring my youth, I enjoyed reading magazine and newspaper articles featuring the elusive "missing link." This catch phrase referred to the singular but not-yet-discovered half-man, half-ape whose fossil would prove the theory of evolution once and for all. It seems that this colorful expression has faded somewhat from common usage. I suspect this is not only because the fossil record has dramatically increased since my youth, but also because evolution is now visualized as a branching tree rather than as the linear progression it was long assumed to be. The fossil record now abounds with transitional life-forms, but for me, picturing some of these fossils as "missing links" is still hard to resist.

Earlier in the twentieth century, scientists and the lay public alike presumed that the authentic missing link between man and ape had been discovered. In 1912, English lawyer and amateur archaeologist Charles Dawson (not to be confused with Charles Darwin) produced skull fragments he claimed to have found near Piltdown, East Sussex. After analysis at the Natural History Museum in London, Dawson's find was hypothesized to be the skull of a 500,000-year-old human ancestor. This specimen became popularly known as Piltdown Man. Although always enveloped in some controversy, it wasn't until 1953 that Piltdown Man was exposed as a fraud: the cranium of a modern but small-brained human, along with an altered orangutan jaw. In 2016, a scientific study using new scientific techniques concluded that it must have been Charles Dawson, himself, who was responsible for assembling this hoax. Some think that it was Dawson's desire to be inducted into the Royal Society that may have motivated him to make this "discovery." Indeed, had he not died prematurely—in 1916 at the age of fifty-two—many think he might have become "Sir Charles Dawson." It took a century, but instead, he will now be remembered as the man responsible for one of the greatest frauds in scientific history. To this day, creationists point to Piltdown Man as a reason not to believe in evolution, claiming that scientists can't be trusted—even though it was scientists who exposed the hoax.

Those with any interest in human origins will have heard of a genuine missing link—Lucy, the *Australopithecus afarensis* skeleton discovered in 1974 near Hadar, Ethiopia, by Donald Johanson and his team. Officially designated AL 288-1, Lucy got her common name from the Beatles' song "Lucy in the Sky with Diamonds," which was reportedly playing loudly in Johanson's expedition camp around the time of this spectacular discovery. In his 2009 book, *Lucy's Legacy*, Johanson writes, "Lucy is a 3.2-million-year-old skeleton who has become the spokeswoman for human evolution. She is perhaps the best known and most studied fossil hominid of the twentieth century, the benchmark by which other discoveries of human ancestors are judged." Johanson says we should think of Lucy's species "as a transitional creature between apes and humans." I would point out that while many more hominid species were to evolve and disappear between Lucy and the emergence of *Homo sapien* three million years later, none has captured our imaginations to the degree that Lucy has. Johanson is also quoted as calling Lucy "the ape that stood up."

Being upright is a bigger accomplishment than it seems at first glance. In order to be completely bipedal—instead of mostly a knuckle-walker like the chimpanzee—our Australopithecine ancestors needed to evolve not only a human-like pelvis, but also a centrally-located foramen magnum, which is the opening through which the spinal column passes. This foramen in chimps and other apes (as well as in all other mammals who are not bipedal) is, by necessity, located toward the rear of the skull. (For proof of this, get down on your hands and knees and see how difficult it would be—given your head position—for you to continually look up to see where you're crawling.)

Lucy was a mere 3 feet, 7 inches tall, with considerable body hair, and she was topped off with a brain of only 450 cubic centimeters—just slightly larger than the chimpanzee brain. (By contrast, our brains average about 1,300 cubic centimeters.) Lucy did retain some limb characteristics that would have helped her climb, but she walked much like we do. Australopithecines similar to Lucy inhabited Earth from about 3.9 to 2.9 million years ago.

Even older than Lucy is the 4.4-million-year-old hominid *Ardipithecus ramidus*. "Ardi," as this Ethiopian fossil is known, was likely one of Lucy's ancestors. We know from its pelvic structure that Ardi would, like Lucy, have been a biped. However, when reading John Gurche's extraordinary book *Lost Anatomies: The Evolution of the Human Form*, and viewing his rendering of *Ardipithecus ramidus*, two things jumped right off the page at me. In addition to gargantuan arms—even longer than Lucy's—Ardi's feet do not look human. Ardi's big toes are markedly abducted (they each point toward the other foot) leaving generous spaces between the first and second toes. These hand-like feet would have made Ardi comfortable swinging through the trees (no doubt helpful in evading predators). With a brain even smaller than Lucy's and extremities made for climbing and swinging, one could make the case that Ardi was more ape than human. Considering that Ardi lived about a million years before Lucy, this makes evolutionary sense.

Genetic mutations have provided frequent new life-forms during most of the history of life on our Earth. However, because so few living things fossilize, the vast majority of these new life-forms will remain missing in action. Sometimes we get lucky, as we did with my favorite "missing link" (no longer missing, of course). For this one it's necessary to journey back much further in time than we did for either Ardi or Lucy. The fossil record tells us that by 360 million years ago, there were creatures similar to modern-day amphibians (salamanders and frogs). However, the only vertebrates that have been unearthed from before 380 million years ago are non-tetrapods (fish). Some years ago, Neil Shubin, a Harvard-trained paleontologist and anatomist who teaches at the University of Chicago, started searching rocks that were 375 million years old hoping to find the fossil of a life-form that transitioned between fish and amphibians. In 2004, after several unsuccessful trips, Shubin and his team discovered a fossil in the high Arctic of Northern Canada. This animal had gills and fins—like a fish—but possessed a neck and a crocodile-like flattened head. What's more, on further study, this 375-million-year-old fossil was found to have small appendages that are part fin and part limb; inside these extremities reside the bony blueprint for all subsequent arms and legs, including ours. Named *Tiktaalik* (Inuit for large freshwater fish), this fish could have propped itself up on the stream bottom and because of its flexible wrists could accomplish primitive "push-ups." Shubin has detailed the discovery and significance of *Tiktaalik* in his compelling book *Your Inner Fish*, and in his PBS documentary of the same name. He says, "Here was an animal that Darwin had predicted: a real anatomical mixture." It's not hard to understand the significance of this fossil. When Shubin took the fossil to his son's preschool show-and-tell, one child asked if it was a fish or a crocodile. Another child then spoke up saying that it could be both!

When thinking about human evolution and all the hominids that preceded us, I can't help but wonder how "missing links" like our diminutive Lucy managed to stay alive long enough to reproduce—much less hang around for a million years. An incredible 1976 find in northern Tanzania—the Laetoli fossilized footprints dated at 3.6 to 3.5 million years ago—is of two fully bipedal, Lucy-like individuals, one larger than the other. They are walking side-by-side with distinctly human strides, and they are close enough to one another to be touching. One set of prints is larger than the other, leading to speculation that they were left by an adult and child, or perhaps, a male with his smaller female partner. These little beings would have faced unimaginable danger every day of their lives; let's hope that they were holding hands.

My Ancestors:
A Small Big History

All that tread
The globe are but a handful to the tribes
That slumber in its bosom.
—William Cullen Bryant, Thanatopsis

*W*ho were my ancestors? How did they live? How did they die? Now that genealogy is almost a fad, these questions seem to interest many. Today, barring certain problems, it is not difficult to discover the inhabitants of your family tree several generations back. For some of us, however, this kind of exploration seems like it should be just the beginning. We're drawn to something bigger—something older.

A Seminal Article

Years ago—and long before my interest in genealogy—I encountered a selection featured as a "Special Article" in the January 1985 issue of the *New England Journal of Medicine* (*NEJM*), titled "Paleolithic Nutrition: A Consideration of Its Nature and Current Implications." The authors were S. Boyd Eaton, M.D., and Melvin Konner, Ph.D., both from Emory University in Atlanta, Georgia. This impressive merger of anthropology and medicine was offered to *NEJM* readers along with the usual more conventional medical research and education papers.

Remembering that it was 1985, there were—at least for me—several stunning observations in this treatise. The authors told us that ". . . early European *Homo sapiens sapiens*, who enjoyed an abundance of animal protein 30,000 years ago, were an average of six inches taller than their descendants who lived after the development of farming." They also added that only now are we "nearly as tall as were the first biologically modern human beings." The authors went on to explain that the "nutritional quality" of the meat—wild deer, bison, horses, and mammoths—from which the "paleolithic population obtained their animal protein . . . differs considerably from that of meat available in the modern [1985] American supermarket." The authors shared a survey of 15 different species of free-living African herbivores, which revealed not only that the African herbivore has a lower fat content than domestically raised meat, but also that its fat is more polyunsaturated and contains certain fatty acids that may even be antiatherogenic, helping prevent the formation of additional fats. The authors reminded us that early human beings also consumed many species of wild plants such as roots, beans, nuts, tubers, and fruits. (Here I should note that many modern fruits and vegetables, highly selected over the last few thousand years for taste, appearance, etc., would have been almost unrecognizable to our Paleolithic human ancestors.) Commenting on observations of "the few surviving hunter-gatherer populations whose way of life and eating habits most closely resemble those of preagricultural human beings," the authors pointed out that members of these technologically primitive cultures (i.e., contemporary hunter-gatherers) who live to the age of sixty or more remain relatively free of coronary heart disease, hypertension, diabetes, and some types of cancer.

Gobsmacked

The fact that modern humans eat a suboptimal diet has not been breaking news for many, many years. However, the point that the diet my ancestors have consumed for the last 5,000 to 10,000 years is, in fact, a relatively new diet for humans—and quite different from our nourishment for the previous 200,000-plus years—was something I had just not considered. Another "aha" moment for me—remember that this was 1985—was recognizing that "the range of diets available to preagricultural human beings" is "the nutrition for which human beings are in essence genetically programmed." What really gobsmacked me back in 1984, however, was learning that—in some ways—Paleolithic humans ("cavemen") were healthier and more robust than their farmer descendants. Of course, present-day writers such as David Christian (*Maps of Time*) and Yuval Noah Harari (*Sapiens*) have confirmed as much in their superb "big history" books. Whether or not these preagricultural hunter-gatherers were also happier than their horticultural descendants remains a thornier question to answer. Since the beginning of recorded history, though, it seems that we have been inclined to dismiss Paleolithic life as miserable and have been taught to feel grateful for civilization. The English philosopher Thomas Hobbes (1588–1679) has often been quoted to portray what life must have been like for early man: ". . . no arts; no letters; no society; and which is worst of all, continual fear, and danger of violent death; and the life of man, solitary, poor, nasty, brutish, and short." Much earlier, Aristotle is quoted to the effect that, "At his best man is the noblest of all animals; separated from law and justice he is the worst."

Out of Africa

Paleontologists and other scientists dealing with early man now agree that anatomically modern humans—*Homo sapiens*—evolved in Africa (our garden of Eden) 200,000-plus years ago. By 60,000 to 80,000 years ago, these humans had reached full behavioral modernity, making more sophisticated tools and creating considerably more art. Around this time, for reasons not completely understood, a small group of these humans living in eastern Africa (possibly as few as one hundred individuals) migrated to Eurasia. Perhaps they were following herds of animals or responding to climate changes or other environmental problems. Within a short time, climatic conditions in northern Africa became dryer, expanding the Sahara Desert and making it more difficult for this group to turn around. Many researchers believe most people today with non-African ancestors are descended from these travelers. Technically, of course, we're all African because *Homo sapiens* evolved in Africa, and we're all descended from those early *Homo sapiens*. However, when we use the term "African" for someone today, it describes a person descended from the many people who didn't migrate (people who are now called "Black"). There is now also evidence of even earlier *Homo sapien* migrations out of Africa, but genetic studies still point to the sojourners in that small migration about 70,000 years ago as the ancestors of most people who are non-African. It also turns out that leaving Africa was not unique to *Homo sapiens*. It's known from fossil discoveries that an earlier hominid *Homo erectus* made it to southern Asia two million years ago!

So, what were these ancestors like who walked out of Africa so long ago and eventually populated the rest of the globe? Prehistory is bound to embody some speculation, but it's quite certain that these early humans would have lived and traveled in small bands of 60 to 100 men, women, and children. In *Genesis: The Deep Origin of Societies*, the eminent biologist Edward O. Wilson tells us that these bands consisted largely of kin and would have been "linked to other bands by kin ties and marriages." He goes on to say, "They were loyal to the aggregate of bands as a whole, although never so much as to preclude murder and revenge raids now and then. They tended to be suspicious, fearful, and occasionally hostile to other communities of bands. Lethal violence was a commonplace." Obviously, this was not paradise! However, anthropologists tell us that hunter-gatherers probably worked less than later agriculturalists and had more leisure time for story-telling, dancing, and what today I suppose would be termed "just hanging out."

It's naïve to think that these clans would have lacked a recurring figure in other primate societies—the dominant male. Still, it's common for paleontologists and anthropologists to use the descriptor "egalitarian" for hunter-gatherer societies, alluding to a certain equality among the clan members. Pharaohs, bishops, or princes—those who do not contribute to the daily work needed for survival—would not have inhabited these clans. It's likely that even being a shaman or a cave painter was just part-time work. Much, much later in our history (1600s), the more egalitarian Native American societies were so attractive to some European immigrants that local governments enacted very punitive laws to prevent more of their male citizens from deserting Puritan society to live with the natives. Native American tribes had no similar difficulty with desertion to the Europeans, even though they coveted some of the immigrants' technology, like metal cooking pots and firearms. However, this egalitarian bent of hunter-gatherer groups has never benefited them when dealing with invaders. Often lacking well-agreed-upon leadership, the Plains Indians—one of North America's last hunter-gatherer societies—suffered grievously when trying to negotiate treaties with the U.S. government—treaties that unfortunately were routinely broken by the government anyway.

Paleolithic Life

Like their *Homo sapien* neighbors who stayed home, the early travelers who left Africa about 70,000 years ago would have been dark-skinned. The gene allowing for pale skin and blue eyes did not appear until 6,000 to 12,000 years ago—20,000 at the most. It evolved first in the area we now consider Scandinavia and later in central and southern Europe. It's assumed that pale skin was a beneficial adaptation to maximize vitamin D synthesis in those humans who took advantage of the retreating glaciers and migrated to northern latitudes—lands that would have had considerably less sunlight than their ancestral African homelands.

By the time this group left Africa, their ancestors had been using fire for protection, warmth, and cooking for a very long time—perhaps up to one million years or more. Any clothing these early travelers wore would have been draped or tied, as bone needles for sewing more fitted garments did not appear until about 20,000 years ago. Their bodies would have been painted and adorned with bracelets and necklaces fashioned out of eggshell beads. For early music-making, they devised flute-like instruments out of bone.

As noted earlier, preagricultural humans would have been adept at making and using stone tools, and they survived by hunting and foraging. They hunted with stone-tipped spears with which they poked, jabbed, and hurled at their prey. The atlatl—a primitive device that greatly increases the velocity and distance of the thrown spear—did not appear until about 31,000 years ago (at the earliest), followed by the bow and arrow some unknown years later. There is disagreement about when Paleolithic humans domesticated the wolf to give us our new best friend. Some believe it may have happened in certain places as early as 40,000 years ago. If, as others believe, this taming happened only about 14,000 years ago, it would still have been in time for our canine friends to be hunting partners and protectors to the humans that migrated to the Americas. These migrants, who walked from Asia across the Bering land bridge, surely needed help as they encountered—and possibly hunted to extinction—mammals as formidable as the huge wooly mammoth. (Recent evidence now suggests that some humans arrived in America before the emergence of the Bering land bridge.)

As early *Homo sapiens* spread across the globe, they would, at times, have drifted, paddled, or sailed on rafts. However, because so much ocean water in this era was taken up by glaciers—drastically lowering sea levels everywhere on earth—early humans could have walked or sailed to places that would now require much longer and more dangerous voyages. Walking—as well as running—for hours and hours at a time is something these folks were just plain evolved to do. Instead of riding them, these early humans hunted and ate horses. It was merely 5,000 to 6,000 years ago that horses were first domesticated, and—along with the wheel—used for transportation!

It's tempting to think of these Paleolithic humans as rather primitive, but they possessed modern-sized brains protected by skulls that had lost the sloping forehead and prominent brow ridges of their more ape-like predecessors. Tall and robust with straight posture unaffected by prolonged sitting and bending over paperwork or screens, some of these low-body-fat hunters would have resembled modern Olympic athletes. In fact, later historical accounts left by European explorers and settlers include descriptions of indigenous people they encountered as having physiques that left the Europeans in awe. By the time they left Africa, one of these humans, properly groomed, and clothed in the latest attire, would not look out of place in a present-day Wall Street firm.

Because of a healthy low-sugar diet, their teeth would have been largely free of dental cavities—a malady of the distant future. Early humans certainly would have fallen ill to certain infections and would have been victims of horrendous trauma, but because of their small numbers and lack of close exposure to domesticated animals (the source of many later infections in humans), many afflictions—especially devastating epidemic illnesses—were far in the future. In fact, the first epidemic—possibly typhus—may not have occurred among humans until 429 BCE in Athens, Greece—many thousands of years later. Camping by lakes and rivers, Paleolithic humans had access to clean, unpolluted water, the scarcity of which would become deadly for their agricultural descendants, necessitating the invention of beer, wine, etc. It's also unlikely that hunter-gatherers suffered from food shortages. Famines were also far in the future—and only after large numbers of people had become dependent on harvested crops.

There's an often-used saying in paleontology that "language doesn't fossilize," illustrating the difficulty of uncovering the origins of human speech. While the hyoid bone in the neck, necessary for speech, fossilizes poorly, a modern-like hyoid bone has been found in the skeleton of a hominid who lived 600,000 years ago. This would imply that speech has been a possibility for a long time. Research on the human genome—responsible for much that is now known about human origins—is also helpful when considering the evolution of language. *Homo sapiens*, it turns out, have two changes in the *FOXP2* gene—thought to be necessary for speech—that other mammals do not have. I can see these early humans with their big brains—armed with the right genes and the requisite anatomy—jabbering to one another—telling stories, planning the hunt, and warning their youngsters about the dangers around them.

Meeting the Neanderthals

As they left Africa, *Homo sapiens* encountered strangers we now call Neanderthals. *Homo neanderthalensis* had inhabited parts of Asia and Europe for a very long time (they lived from approximately 450,000 years ago until about 30,000 years ago). Because of advances in genetics, it's now understood that most of us who are descended from the "out-of-Africa" *Homo sapiens* can also count Neanderthals as ancestors. (In my case, almost two percent of my genome is Neanderthal.) While it may not have been "love at first sight," some of these contacts were obviously more than a mere how-de-do. Although they were shorter and stockier than *Homo sapiens*, Neanderthals were not the brutes they've often been portrayed to be. In his Great Courses lectures, "The Rise of Humans," Professor John Hawks tells us that there's evidence that later Neanderthals painted objects and themselves, showed compassion for the ill and injured, and buried their dead. Dr. Hawks says that Neanderthal skeletons can show injury patterns similar to that seen in modern-day rodeo cowboys. This leads me to think that Neanderthals must have been fearless hunters who made their kills at close quarters. As with other early hominids, it's difficult to ascertain how verbal Neanderthals were. However, they had the same *FOXP2* genetic mutations we have, and a similar hyoid bone (however, only one specimen has been recovered). In addition, like Homo sapiens (and unlike, for example, chimpanzees), Neanderthal middle ear bones are better attuned to hearing the frequencies necessary for speech discrimination.

Farmers and Herders

Many think the transition from hunter-gatherer to farmer started roughly 12,000 years ago between the Tigris and the Euphrates rivers—in what we now call the "Fertile Crescent." (Mesopotamia literally means "the land between the rivers.") This agricultural (or Neolithic) revolution, aided by milder interglacial temperatures, led to much larger human populations, city-states, and governments. Then came the written word, making possible *The Epic of Gilgamesh* in 2100 BCE—perhaps the beginning of the arts, literature, and humanities. In fact, it seems that almost everything we now hold dear would not have been possible without farming and the domestication of animals.

You don't have to believe in reincarnation to wonder how you might have fared—even just survived—as one of your many ancestors. Rather than risk getting on the wrong side of a medieval English king, only to be hanged, drawn, and quartered, some might prefer taking their chances as a cave-painter 20,000 years earlier. Being a gifted genius has never been safe, either. Less than 500 years ago, the brilliant forty-two-year-old Michael Servetus—physician, humanist, theologian, cartographer, and first European to correctly describe the pulmonary circulation—was burnt alive atop a pyre of his own books. His crime? Disagreeing with religious authorities in Calvinist Geneva over theological questions such as predestination and The Trinity.

We tend to think that life improved for humans as civilization advanced. But did it? Not for everyone. In *Maps of Time: An Introduction to Big History*, David Christian tells us: "In 1800 in England the adult male members of the titled nobility stood a full five inches taller than the population as a whole."

Suffering

I suspect that for most of history—regardless of their class or wealth—suffering was an inescapable part of life for our ancestors. In 1811, "Nabby" Adams (Abigail Adams Smith), beloved daughter of John and Abigail Adams, consulted respected New England surgeon John Warren for treatment of an obviously malignant breast tumor. Employing state-of-the-art medical care, Dr. Warren performed a mastectomy in an upstairs bedroom of the Adams family home with Nabby strapped to a chair. After Dr. Warren sliced off Nabby's breast with a wooden-handled razor and excised some lymph nodes from her armpit, he pressed a hot iron spatula, fresh out of a nearby oven, to her chest wall to cauterize the bleeding. All of this—plus suturing—without the benefit of anesthesia. Can you imagine? Unfortunately, it may be that until recently, excruciating pain could be an accepted fact of life—and death. In 1842, twenty-six-year-old John Thoreau, Jr., older brother of author Henry David Thoreau, died in Henry's arms of tetanus—known then simply as lockjaw. Years before any effective treatment, this would have been an agonizing death. Untreated tetanus causes an exaggerated arching of the body along with muscle spasms for days that are severe enough to tear muscles—even to fracture large bones.

Personally, I don't yearn to walk in the footsteps of any of my ancestors—Paleolithic or Neolithic—and certainly none who lived before acceptance of the germ theory. While I will try to eat a little more like my hunter-gatherer forefathers did, I will remain grateful for the fruits of the agricultural revolution. It's unlikely that I will be eaten by a large predator, be persecuted for heresy, or die from tetanus. My major surgeries, thankfully, were not torture by another name.

A Steep Price

It's impossible for many of us to ignore the steep price that's been paid for many of our Neolithic advances, as marvelous as they've been. When my grandparents were born, the fastest mode of transportation was the train—no automobiles, no airplanes—but pollution and air quality were only problems in certain industrial centers—and the oceans were not full of plastic debris. We've split the atom, but we may not be responsible enough to control the products of that fission. I can't stop wondering—how will our descendants remember their ancestors?

Trusting History

History is what we choose to remember,
and we have no alternative but
to do our choosing now.
—Joseph Ellis, American Dialogue

istorical commentaries—books, podcasts, and documentaries—can be fascinating. But how much should we trust them? Sayings such as "to the victor go the spoils" and "history is a puzzle with most of the pieces missing" ought to caution us as we digest histories, both oral and written.

Even historians researching recent historical events can find it challenging to uncover the truth. When he began his research on the life and times of President Lyndon Johnson, renowned historian and author Robert Caro suspected he was missing important oral histories. This prompted Caro and his wife to relocate from New York

City to Texas. By living among the folks whose recollections and stories he needed, Caro was able to gain their confidence and trust. It was only then that Caro began hearing the more confidential and accurate remembrances of Johnson's formative years—remembrances that had earlier been denied to him as a visitor.

Unlike Caro's situation, much of the history that interests us concerns events further in the past, whose subjects and witnesses are long dead and gone. In recreating these characters' lives, loves, and battles, historians are sometimes fortunate to gain access to fresh material such as long-missing or overlooked letters, diaries, and other documents. How lucky it was for history buffs when David McCullough—another Pulitzer Prize-winning author—learned about a trove of such treasures at Marietta College in Marietta, Ohio. This discovery prompted him to write *The Pioneers: The Heroic Story of the Settlers Who Brought the American Ideal West*. But, more often than not, those who are writing about the distant past stand chiefly on the shoulders of earlier historians. All history is, to some extent, revisionist history. However, few would be content with reading and discussing only the accounts of ancient scribes. Each generation welcomes the interpretations of history written by its contemporaries—renditions that hopefully have benefited from research techniques or tools not available to past historians.

There is no better example of historical controversy meeting modern research than that surrounding the last Plantagenet king, Richard III of England (1452–1485). For many of us, hearing Richard's name conjures up an image from Shakespeare's famous play of the evil, deformed king desperately wandering the battlefield exclaiming, "A horse, a horse! My kingdom for a horse!" Richard—the last English king to die in battle—did perish at Bosworth Field on August 22, 1485, when his army was defeated by Henry Tudor's forces. This is the same Henry who became Henry VII (the first Tudor king of England and father of the more famous—or infamous—Henry VIII). It has long been held by some that the hateful Richard was a fabrication of Tudor historians, who were writing a version of history guaranteed to please their Tudor kings. In fact, it was only a hundred years later when William Shakespeare wrote his splendid play depicting a nefarious hunchback king who murdered his defenseless nephews—the princes in the tower—and claimed the English crown for himself.

There is now a Richard III Society with over 3,500 members worldwide dedicated to the rehabilitation of Richard's reputation. Members of the society believe that Richard was a just and chivalrous king whom Shakespeare—basing his portrait of Richard on those narratives left by Tudor historians—horribly maligned. It's generally accepted that after the battle in which he perished, the slain Richard's naked corpse was unceremoniously flung over a horse and taken from Bosworth Field for a quick burial. Using ancient maps and documents, society members deduced that a logical site for Richard's hurried burial would have been a nearby chapel—a chapel whose archeological remains lay hidden under what is today a carpark in present-day Leicester, England.

Based on this twenty-first century detective work, enough money was raised to close the carpark for two weeks and begin excavation. To the delight of interested onlookers, bones were soon discovered in what the archeologists determined would have been the friary of the ancient chapel. Excitement really began to mount when it was observed that the spine of this skeleton was scoliotic (characterized by a spinal curvature). Next came weeks and months of investigation by experts in various scientific disciplines. With the help of carbon dating and forensic pathology—including comparing the injury of the skeleton to that described in ancient chronicles—it became possible to hope that Richard's bones really were those that had rested under the carpark. Researchers then found a direct descendant of Richard's sister who was willing to have his DNA compared to that salvaged from the excavated skeleton. In what must have been the end of a tremendous amount of breath-holding, a positive DNA match was announced. Richard III had been found!

So, what was learned about the historical Richard? He did have a scoliosis significant enough that he would have carried one of his shoulders somewhat higher than the other, but this abnormality may not have been readily apparent when Richard was fully clothed. He also did not have the withered arm that Shakespeare bequeathed to him. Richard, it seems, would not have appeared horribly deformed to his subjects or to his enemies.

Made possible by a mostly intact skull, the final step in this quest was facial reconstruction—what did Richard look like? The reconstructed face has been judged as rather handsome, but somewhat gracile, or feminine. Here, it is interesting to note that earlier in the investigation, an archaeologist—as she was freeing his bones from the earth—commented on an aspect of the skeleton's pelvic anatomy that was often seen in females. She then reassured worried onlookers that this pelvic anatomy could also be seen in males. This physiognomy may be compatible with a surviving narrative from Richard's contemporaries, who remarked that they were surprised by his effectiveness as a fighter given his feminine appearance! Identifying and describing Richard's skeleton—even knowing what he looked like—doesn't help us determine how evil or good he was. However, we now know that Shakespeare exaggerated Richard's deformities. Did he do the same with his character?

When Richard's older brother Edward IV died, Edward's son—twelve-year-old Edward V—succeeded his father. Richard, who had always been loyal to his brother, was named Lord Protector of the Realm. Using this office, Richard succeeded in having the new king and his even younger brother declared bastards, then proceeded to seize the throne for himself. The young princes, who had been living in the Tower of London—supposedly for their own protection—were never seen again. This certainly sounds damning, but it was only after Richard's death that he was accused of having the boys murdered, and some have instead blamed the Tudors for the boys' disappearance.

In the Middle Ages, it seems that betrayal and treason were everyday behaviors. Sons rebelled against fathers, and brother fought against brother. It was not unusual for the combatants to reconcile, only to betray one another again. Power was paramount! One did anything needed to gain power and to keep that power. Nowadays, we like to romanticize these medieval rulers and aristocratic families ("In days of old when knights were bold . . ."). However, a closer look at these folks—their treachery, brutality, and "might-makes-right mindset"—can be sobering. Rather than gallant warriors, we're reminded more of modern warlords or crime bosses. Still, living soon after the Dark Ages and well before the Enlightenment, the conduct of these royals was probably just what their peers expected—even admired. Rather than asking if Richard was a villainous king or a just king, a more germane question may simply be whether he really was as awful as the Richard in Shakespeare's play.

Discovering Richard's bones and learning more about his physical appearance tells us nothing about his character. However, the identification of Richard's remains may end up benefiting his historical reputation—whatever the historical truths may be—as more of us are motivated to look beyond Shakespeare's Richard. Meanwhile, history junkies (myself included) will stay tuned, keep reading, keep watching, and gobble up the latest version of the past—missing parts, character assassinations, and all.

Marching Through History Together: Some Thoughts on Population Genetics

*E*veryone understands that for each generation back, the number of potential direct ancestors is doubled. One generation back I have two direct ancestors (my parents), and two generations back I have four direct ancestors (my grandparents). The math so far is obvious, but it gets more interesting. What that doubling effect means is that 40 generations ago—only about 1,000 years—you would have one TRILLION potential direct ancestors!

Most experts estimate a world population 1,000 years ago of, at most, 400 million people. However, in Europe—where almost all of my ancestors then lived—that population would be only 50 million. There are presently more people living on the earth than at any single moment in the past. It would seem that there are simply not enough ancestors to go around! Historically, about one-third of people do not have children, which shrinks the ancestor pool even more.

What becomes obvious from looking at these numbers is that our family "tree" must be populated with the same people over and over on different branches of each generation. In other words, you are descended from the same individuals many times over. An example of this can be found on my wife's family tree. Her paternal grandfather's parents are directly descended from two different sons of a man who lived from 1629 to 1702. This common ancestor is her grandfather's third-great-grandfather on his father's side of the family and his fourth-great-grandfather on his mother's side. Therefore, this pair of paternal great-grandparents were third cousins once removed.

For most of us this means that we would not have to go back many generations to find common ancestors with many of our friends and acquaintances. Proceeding further back, we would discover common ancestors with people now living all over the world. We all have lots of cousins! I find it helpful to envision our actual family tree as being more diamond-shaped than it is tree-shaped.

A number of years ago, I was busy researching my own ancestry. Naive to the above ideas, I was startled to find a fifteenth-century common ancestor for my maternal grandparents. That meant my mother's parents were—unbeknown to them—tenth cousins once removed. By the time I began researching my wife's family tree, I was prepared to discover that we might be distant cousins. What I wasn't prepared for was the eerie feeling that our families have marched together through history.

I found that we both could be descended from Norman knights who helped William the Bastard become William the Conqueror in 1066. Later, in 1215, distant grandfathers from both trees were among the 25 barons who forced King John to sign the Magna Carta. I saw mention of a marriage in the 1300s between possible ancestors, and then I discovered a more recent common ancestor, Bygod Eggleston (1586–1674). Bygod's great-grandfather was Sir Francis Bigod (1507–1537) who fomented a revolt against Henry VIII that is still remembered today as Bigod's Rebellion.

It's important, however, to remember that during ancient and medieval times, most of our British ancestors would not have been knights or barons. Many would have been serfs, peasants, and others just scraping out harsh, menial lives. We know that both my wife and I have ancestors who emigrated from the same districts in England. Over the several-hundred preceding years, some of these immigrants' ancestors would have worked side by side, worshipped together, and suffered together. Some would have married. Unlike the knights and barons, these ancestors' lives went unrecorded and unremembered—but they are still our grandfathers and grandmothers.

Leaving England did not mean the end of this shared history. In 1620, my wife's tenth-great-grandfather and my eighth-great-grandfather risked an ocean crossing together on the *Mayflower*. Each of us has distant grandparents who, in 1637, were banished from Massachusetts Colony with Anne Hutchinson, becoming some of the earliest settlers of Rhode Island. Both my wife and I have numerous immigrant ancestors who called New England their new home in the 1600s. That early in colonial times, the population in New England was small enough that our ancestors would surely have rubbed shoulders.

My wife's last ancestor to leave England was a paternal great-grandfather, who sailed from England in October, 1866. My last ancestor to leave England was also a paternal great-grandfather. My great-grandfather sailed for New York in—you guessed it—October, 1866, but on a different ship than her great-grandfather did. (One was from Cambridgeshire, and the other was from Yorkshire, so it's unlikely they ever met.)

In light of our possible common ancestors, my wife and I are anywhere from eighth to tenth cousins—not nearly close enough to raise any eyebrows. Finishing this project, I feel sure that there are many yet-to-be-discovered connections between our trees. (Consider all those unrecorded lives!)

The 800-plus ancestors on her tree and the 600-plus on mine represent only a small fraction of our ancestors. Even going eight to ten generations back would mean several thousand ancestors, not several hundred; some of those thousands would have been shared ancestors. We are all more closely related than most of us could have imagined!

The American Dream

Give me your tired, your poor,
Your huddled masses yearning to breathe free,
—Emma Lazarus, "The New Colossus"

An oft-used phrase—now a maxim—the term "the American dream" has come to convey the unlimited opportunities thought to be available to people living in the United States of America.

After becoming interested in genealogy, I discovered—to my surprise—that most of my immigrant ancestors, as well as those of my wife, landed on American shores during the 1600s and 1700s. Although these early arrivers would have faced hardship, danger, and death, they were often escaping conditions in Europe that they considered to be intolerable. After surviving risky voyages, many found their new lives largely improved. A good example is John Rathbone (1629–1702), my wife's seventh-great-grandfather.

Civil unrest and poor economic conditions as well as an outbreak of plague may have prompted John and his new wife, Margaret, to leave their English home for America in 1649. In Lancashire, England, John and his crowded family would have lived in a small, two-room, one-story cottage with a packed-dirt floor and windows lacking glass. When he died, this illiterate immigrant had become a prominent citizen of Rhode Island Colony and one of the largest land owners on Block Island. He also owned a home and a business in Newport.

Although many of my early ancestors may have prospered in some way, emigrating to the New World has never been a guarantor of success or wealth. Consider my third-great-grandfather John Baars (1755–1824). Unlike many of my earlier colonial ancestors, John was not an English Puritan; rather, he emigrated from Hanover, in today's Germany. We know that by January 1781, he was living in Massachusetts, where he enlisted in the Continental Army, serving in the Sixth Massachusetts Regiment for three years. We also happen to know something of his later life because of an extant pension request. In 1820, he was granted a pension of eight dollars a month retroactive to 1818. The physicians who examined him concluded their statement by indicating they were familiar with several pensioners and "think his necessity is as great as any we know and*****(?unless) he receives a pension he

will be unable to support his family but for a short time."
John was a widower with five children. Among his possessions listed are several pieces of used and broken-down
furniture, kitchen items, a few farm tools, three shirts, two
calves, four pigs, and what he describes as probably enough
grain for his family until the next harvest. This immigrant
was not living anyone's idea of the American dream.

Many of my (and my wife's) early immigrant ancestors were Puritans who—while they wished to prosper
economically—left England chiefly for religious reasons.
In Massachusetts, these folks hoped to live in a snug little
Puritan theocracy without interference from the Church
of England. For them, forcing their neighbors to also live
according to their biblically rigid norms (falling asleep
in church could mean jail time) was their version of the
American dream. However, we also both had ancestors
who chafed at this and were expelled from Massachusetts,
only to try and attain a slightly different version of the
American dream in Rhode Island or elsewhere. They
might have agreed with H. L. Mencken, who much later
famously wrote: "Puritanism—the haunting fear that
someone, somewhere, may be happy."

The last two immigrant ancestors on my family tree—
my paternal great-grandfathers—were, it seems to me,
near-perfect embodiments of our so-called American
dream. Reuben Dickinson and Benjamin Pont both

lived in small English Cambridgeshire villages where their families were nonlandowning laborers occupying the bottom rung of the Victorian economic ladder. With a British class system that allowed almost no chance for improvement, emigration—with all its uncertainties and hardships—would have offered an opportunity like no other. As will become apparent, both of these young men, despite being poor, were in many ways prepared to succeed in their new homeland.

On October 20, 1866, sixteen-year-old Benjamin and his older brother sailed from Liverpool on the *St. Marks*. Although steamships were common by this time, the brothers truly were sailing. The *St. Mark* was an American sailing ship (a freighter) carrying cargo plus 50 passengers who paid 3 pounds and 10 shillings, or about 17 dollars, each for the 49-day trip. As part of their fare, the passengers were given weekly rations, which they cooked and prepared themselves. After buying their tickets, the brothers were left with a total of 11 English shillings to their name. This meant waiting in Manhattan at Castle Garden (America's first immigration center) upon arrival for the 11 days it took them to contact their uncle and receive the railroad fare he sent for their trip to Annawan, Illinois. Having an uncle who had already prospered in America was one of those blessings that allowed my great-grandfather to access the American dream. Benjamin and his brother lived in

Illinois for about five years before leaving to homestead in eastern Nebraska with yet another uncle. As the story goes, this uncle had barely escaped England—and deportation to Australia—after an altercation in an English tavern where he had offended some of the landed gentry. One of the things that I'm sure Benjamin and his brother gladly left behind in England was their illegitimacy. Their mother, Mary, completed the space for "father" on both of their birth certificates as "unknown." Impoverished and illegitimate, immigrating to America may not have been a difficult decision for Benjamin and his brother.

Reuben's emigration from England in May of 1864 at age eighteen was somewhat different from Benjamin's sojourn. There is more than one version of Reuben's trip to America, but the most credible one is also the most interesting. Like Benjamin, Reuben was a member of the lower class; he was working as a coal miner in England at the young age of sixteen. The foreman of the mine had arranged to send his own son to the U.S. to work on the railroad as a section hand laying track. The contract was for three years' room and board and $100 to be issued to him at the end of his contract when he returned to England. At the last minute, the foreman's son came down with the "fever," and Reuben got to go in his place. Reuben did work for the railroad for a year but then quit. He was now in the U.S. illegally but was rehired "down the way" with pay. In Annawan, Illinois, Reuben met Alice, a young woman who worked in either a restaurant or a laundry. They were married in December of 1865, and in 1873 they traveled with their three children in a crowded oxcart to their homestead in Nebraska.

I assume that Reuben and Benjamin either knew one another in England or that their families were acquainted in some way. Both seemed to have connections with Annawan, Illinois, and ended up with farms only one mile apart in Nebraska. It was there that my grandparents—Benjamin's son Homer and Reuben's daughter Alice—grew up as neighbors and in 1904 were married.

While I've proposed that Reuben and Benjamin lived the American dream, it may have seemed more like a bad dream to both for some time. After bringing his family west in the oxcart and homesteading on his 80 acres, Reuben also worked for the railroad in Nebraska. Leaving Alice to look after the homestead and their growing family all week, Reuben would walk to the farm Saturday night and back to his railroad job Sunday night. This was a round-trip journey of about forty miles! We know they were still living in a sod house in 1882. In spite of the challenges, within six years Reuben was able to buy a 260-acre farm and eventually become a pillar of the community. From his biography in *Nebraska History*, we learn that "Mr. Dickinson is a prosperous man of affairs, and owns 260 acres of fine farmland, aside from good property in Schuyler." His biography concludes by saying, "He is widely and favorably known." In 1901—37 years after he left his boyhood home—Reuben returned to England on a 100-day vacation. Upon his return, the local newspaper reported that Reuben "says that he comes home better than ever pleased with this country."

I was struck by how completely Reuben had left his immigrant identity behind and had woven himself and his family into the fabric of America when I read my grandmother's obituary. Reuben's daughter Alice died in 1929 at age forty-five and was described as a "daughter of the prairie" from "pioneer parents."

Although Benjamin, like Reuben, also escaped the poverty and social class restrictions of his English homeland to find success on the Nebraska prairie, he struggled for years with blizzards, grasshopper infestations, prairie fires, and near famine. Benjamin persevered, and later in his life, he was described in the *Compendium of History, Reminiscence, and Biography of Nebraska* as "one of the most respected citizens in Stanton County."

Even though they were poor, in some ways Benjamin and Reuben arrived in the United States armed for success. Both, of course, spoke English, but because of recent educational changes in England, they would have been among the first of their socioeconomic class who could both read and write in their native tongue. We learn from Benjamin's obituary that he even used this literacy to help his non-English speaking neighbors who had come from the "Old Country . . . assisting them to learn the language of America and befriending them in many ways." But, what may have been most crucial to their success was that—like many immigrants who knew they had no real prospects to return to—Reuben and Benjamin emigrated expecting to work very, very hard, which they certainly did.

Many of us have ancestors who lived this American dream, but regrettably, others' ancestors instead suffered through an American nightmare. Whether they came to America in 1620 or 200 years later, immigrants invariably lived on land taken—usually by force—from Native Americans whose ancestors may have been there for at least 13,000 years. Besides what must have seemed like an unlimited number of settlers hankering for land, European ships brought disease and technology that was ultimately overwhelming for the native populations. On the other hand, enslaved Africans—if they survived the trip—lived an American nightmare from the moment they landed! It's sobering, if not heartbreaking, to realize that my ancestors who were fortunate enough to live the American dream did so at a cost so devastating to others.

When I discovered how successful and prosperous two of my great-grandfathers—Benjamin and Reuben—were, I was stunned! I was a grandchild of Reuben's daughter and Benjamin's son, but growing up, I had grandparents and parents who were barely getting by. Reuben and Benjamin both died before the Great Depression. They didn't live to see the widespread financial ruin of our country, followed by the despair of drought and dust on America's farms. Farms were lost to the bank with no money for college, and no jobs. My grandparents and parents were left to make the best of it, and they did their best. They worked whenever and wherever they could to give my generation our chance at the American dream.

American Exceptionalism or Exceptional Americans?

"Intellectually I know America is no better than any other country; emotionally I know she is better than every other country," the novelist Sinclair Lewis remarked in 1930.

—Jon Meacham, The Soul of America

*A*merican exceptionalism is a belief proudly held by some in our country. For others, though, this notion that residing in the United States makes them in some way superior to or more talented than the citizens of other countries is not only invalid, but also boastful. Those promoting the concept of American exceptionalism invariably point to our Founding Fathers as the first important examples of this concept. Those who find the idea of American exceptionalism wanting might agree only that these early patriots were exceptional people.

How did it happen that men the caliber of Benjamin Franklin, Thomas Jefferson, and John Adams all lived in the remote North American British colonies at just the right time in history to lead an earth-shaking revolution? It turns out there were unique factors at play, both within the colonies and in the Western world as a whole, that help account for this extraordinary happening. To begin with, the colonists in America thrived under the protection of the British Empire, with little worry about conquest by foreign powers such as France or Spain. Additionally, for many years preceding the American Revolution, the British government was preoccupied with European wars. Due to this preoccupation, as well as to the "Big Pond" that separated North America from England, the colonists enjoyed more freedom than either the King or the Parliament probably ever intended. This unintended freedom meant that the colonists had ample opportunities to practice an unprecedented amount of self-government that would later prove invaluable. The American colonists—living in a society without a rigid class system—had governing bodies such as legislatures and courts in place before the Revolution, and they were adamant about using them. One has only to compare the American Revolution with the chaotic and grotesquely bloody French Revolution, which resulted in Napoleon's reign and years of European wars. Historically, outcomes like the French experienced are the norm, while the American outcome is the exception. This successful American Revolution—followed by the peaceful establishment of a democratic republic—no doubt planted the seeds for a belief in American exceptionalism.

The years of benign English neglect were certainly a breeding ground for budding patriots, but they cannot fully explain the emergence of some of the luminaries we remember as our Founding Fathers. It's helpful to understand that at this time in history, the more educated among these men were products of the Enlightenment, or "Age of Reason," which began in Europe shortly after the Pilgrims ventured to Massachusetts. (While products of the Reformation, the Separatists that we now remember as Pilgrims were in no way Enlightenment thinkers.) In addition to studying the classics, Founding Fathers such as Thomas Jefferson and John Adams read deeply from Enlightenment literature that emphasized science, reason, and skepticism. Even as students, they learned to question religious dogma and authority that would have gone largely unchallenged by their ancestors. From the Englishman Francis Bacon (1561–1626), they became aware of empiricism and the scientific method. (Bacon was also a noted Parliamentarian who proposed reforms to English law.) Englishman John Locke (1632–1704) emphasized the equality and independence of people in their natural state, implying consent of the governed. (Scholars believe Locke's philosophy is reflected directly in the American Declaration of Independence.) When enjoying the works of the popular French author Voltaire (1694–1778), certain Founding Fathers would have seen it was possible to question intolerance, established institutions, and religious beliefs. From the seminal work of Sir Isaac Newton, (1642–1726), they came to understand that the earth and the universe obeyed natural laws, not superstition. Some were obviously aware of the French lawyer and philosopher Montesquieu (1689–1755), who—long before the creation of our Constitution—had advocated separation of governmental powers into executive, legislative, and judicial branches. It's also crucial to remember that most American colonists considered themselves to be English, and that for some time, English citizens at home had been gaining significant rights. Although King John quickly ignored it and persuaded the pope to excommunicate some of the barons who forced him to sign it, the Magna Carta (1215 CE) first asserted that Englishmen had certain rights. Later, the venerable English jurist Sir Edward Coke (1552–1634) influenced decisions declaring not only that the king was subject to the law, but also that even acts of parliament could be voided if against common right and reason. Of course, it took centuries for Englishmen everywhere to gain anywhere near the semblance of rights they take for granted today, but British colonists in America believed that they, too—as Englishmen—possessed certain rights. Educated colonists studied English history including the Magna Carta, Edward Coke, and the English Civil War (when Charles I was beheaded), and they understood that kings were far from invincible.

Have we come to the tidy conclusion that it was only Enlightenment-educated colonists who arrived at sufficiently radical notions of their rights as Englishmen to rebel? Consider another Founding Father, Thomas Paine, who is honored by some historians as "The Father of the American Revolution." Paine authored the two most influential pamphlets published at the beginning of the Revolution: *Common Sense* and *The American Crisis*. An estimated 500,000 copies of *Common Sense* were sold, helping his rhetoric to reach most of the two million colonial residents of the time. His stirring words in *The American Crisis*—beginning with: "These are the times that try men's souls: The summer soldier and the sunshine patriot . . ."—were read to George Washington's soldiers at the general's instruction three days before the Battle of Trenton. Washington's daring Christmas crossing of the Delaware River to defeat the British was nigh life-or-death for the Revolution; following a sting of humiliating defeats, this stunning and unexpected victory energized colonists at a time when many, including Washington himself, worried that the rebellion was a lost cause.

Tom Paine was an iconoclast who advocated a more radical democracy than many other Founding Fathers would have found comfortable. Paine even believed that men who were not property owners should have the right to vote. Not one to avoid trouble, Paine later got caught up in the French Revolution, barely escaping with his head. While in France, Paine wrote *The Age of Reason*; this pamphlet applied to organized religion what Paine no doubt considered to be the same logic and reason with which he earlier had so deftly attacked George III and despotism. It turned out that postcolonial Americans were not ready for a deist to challenge the legitimacy and divine inspiration of the Bible, or even to question its historical accuracy. To many Americans of that era, freedom of religion meant the freedom to worship at the organized Christian church of your choice—not to criticize Christianity or the Bible.

After surviving the French Revolution, Paine returned to the United States but was ostracized largely because of those heretical religious beliefs he had espoused so clearly in *The Age of Reason*. Showing unusual courage, Thomas Jefferson—in spite of the public outcry—invited Paine to stay with him for a time in the White House. It seems that Paine was essentially excommunicated from American society; only six people attended his funeral in 1809. Of course, stories about the past are seldom as clean, concise, and straightforward as some of us might like them to be. Paine was irascible, difficult, and—at least late in life—possibly drunk a good deal of the time. In an essay on Paine she wrote for the book, *Revolutionary Founders*, historian Jill Lepore tells us that in 1806: ". . . a neighbor of Paine's came across the old man himself, in a tavern in New York, so drunk and disoriented and unwashed and unkempt that his toenails had grown over his toes, like a bird's claws." Time has been good to Paine; history has treated him more kindly than did his peers. Today, Paine's contribution to American freedom is widely acknowledged.

Those espousing the idea of American exceptionalism may want to avoid including this exceptional Founding Father as another example of their belief. After all, Thomas Paine didn't emigrate from England until 1774—just in time to become "The Father of the American Revolution."

The Greatest American

It is men who wait to be selected,
and not those who seek,
from whom we may always expect
the most efficient service.
—Ulysses S. Grant, Memoirs

*W*ho is the greatest American? After reading Joseph Ellis's splendid book *The Quartet*, I suddenly realized that I would answer that question now—in my eighth decade—as I would have in my first: George Washington.

I'll admit to serious flirtations over the years with other great Americans. Some say that when he penned the Declaration of Independence, Thomas Jefferson "invented" America (or at the least wrote its birth certificate). Shaped by the Enlightenment, Jefferson and his peers John Adams and James Madison were scholars, classicists, and political philosophers on a par with any

well-educated eighteenth-century European. Benjamin
Franklin was America's greatest Enlightenment figure
and for years was the "face" of the American Colonies
in Europe and elsewhere. It's difficult to imagine our
country without the leadership of these great Founding
Fathers. Abraham Lincoln is one of the most admired
figures in American history, but, I wonder: Would there
have been a country to save without Washington? Lincoln
himself said, "Washington, the mightiest name on earth."
(Author Sarah Vowell quipped that in heaping this praise
on Washington, Lincoln was "ignoring slackers like The
Buddha and Jesus Christ.")

James Madison certainly deserves to be remembered as
"The Father of the Constitution." However, Ellis reminds us
that Madison—along with Alexander Hamilton and John
Jay—understood that without Washington's presence the
Constitutional Convention would likely be doomed. Many
state delegates were adamantly opposed to surrendering
any of their hard-won liberties to a strong central government. However, General Washington and his troops had
suffered terribly during the Revolution at the hands of a
weak Continental Congress. No one understood better than
Washington the need for a government stronger than the
one provided by our Articles of Confederation. Answering
the call again, Washington left Mount Vernon and served
as president of the Constitutional Convention. Once the

Constitution was completed, the document still faced tremendous opposition. Widespread understanding that the one-and-only George Washington would serve as the first president under this new constitution may have been essential for its eventual ratification. Ellis calls Washington "the Foundingest Father of them all." Astonishingly, Washington was not only vital to our independence, but, also became the worldwide symbol of liberty in his era.

Why was Washington so revered by his peers? The image most of us have of him today is from portraits later in his life (think dollar bill). But, by all accounts, the general—in his forties, and commanding the Continental Army—was a magnificent human being! We have multiple descriptions of him that attest to his bearing, bravery, and integrity. Taller than most, he had a physically compelling presence. He was a demon on horseback.

Washington's bravery in battle was first noted when he was a young officer in the French and Indian War. As our commanding general in the American Revolution, Washington was known to ride into his retreating troops, exhorting them to turn and charge. After one such battle, a soldier wrote home: "O my Susan! It was a glorious day, and I would not have been absent from it for all the money I ever expect to be worth . . . when I saw [George Washington] brave all the dangers of the field and his important life hanging as it were by a single hair with a thousand deaths flying around him." President James Monroe, who had served under Washington—and, it is claimed, crossed the Delaware with him—admiringly said, "A deportment so firm, so dignified, but yet so modest and composed I have never seen in any other person." Lafayette fought alongside Washington and was forever in awe of him. Late in life, the French hero of our Revolution declared, "I thought then as now that I had never beheld so superb a man."

When Abigail Adams—perhaps our most articulate Founding Mother—first met Washington, she was moved to quote Dryden: "Mark his majestic fabric, he's a temple sacred both by birth and built by hands divine." On another occasion, Abigail observed that Washington, despite his regal appearance, was neither haughty nor overly imperious. She records, "The gentleman soldier look agreeably blended in him. Modesty marks every line and feature in his face."

Though not remotely as well-educated as Jefferson, Adams, or Madison, Washington may have been uncommonly wise. When relinquishing power, he commented that he, "did not fight George III to become George I." In an observation that could give us hope yet today, Washington wrote to Lafayette two years before the Constitutional Convention: "Democratical states must always feel before they can see. It is this that makes their government slow but the people will be right at last."

George Washington's hero was Cinncinatus, the Roman general who returned to his farm after saving Rome. When George III was told that Washington was also going to lay aside power after the Revolution and resume private life, the astonished monarch said, "If he does that, he will be the greatest man in the world." Many would still agree.

For Katie: Reflections on Ron Chernow's *Alexander Hamilton**

The ten-dollar Founding Father without a father.
Got a lot farther by working a lot harder, By
being a lot smarter, By being a self-starter.
—Lin-Manuel Miranda, Hamilton

*A*s you grow older you'll likely have a bank account, borrow money for a mortgage, enjoy the fruits of living in an industrial nation, and feel some sense of security because of a large standing U.S. military force. For all of these things, you can (in part) thank Alexander Hamilton.

This book is not just the biography of a brilliant and important Founding Father. It is also a detailed examination of the politics and history of revolutionary and postrevolutionary America. The author spins a compelling narrative and is a wordsmith worth envying. Early in the book he describes Hamilton as "the messenger from the future we inhabit."

He also tells us: "In all probability, Alexander Hamilton is the foremost political figure in American history who never attained the presidency, yet he probably had a much deeper and more lasting impact than many who did." I would eagerly substitute the word "most" in place of "many."

You will learn that the very young and inexperienced Hamilton was not only a valiant officer, but also became one of General Washington's most trusted aides (most of Washington's field orders that have survived are in Hamilton's handwriting). He later was President Washington's most valued cabinet member—many of whose views the president adopted.

This book illustrates just how difficult it was for Hamilton, Madison, and Washington (as well as others) to get our Constitution written and ratified. Having just been through a war for independence, many colonists were adamant about not yielding their existing state's rights to a strong central government.

Since my high school days, I've been fascinated with the Hamilton vs. Jefferson story in American history. Later, I came to appreciate their battle as a template to better understand not just American history, but even modern politics. Chernow deftly explores this conflict and uses it to help explain the appearance of political parties (mentioned nowhere in the Constitution) so early in our history. While reading this book, it's impossible not to watch the seeds of the American Civil War—erupting over a half-century later—being planted in postrevolutionary America.

In view of the acrimony of modern day politics, it was strangely comforting for me to see that even in early America, politics were nasty. It was also reassuring to note that then, as now, politicians were able to discard ideology they professed as candidates in order to intelligently govern. For example, when Jefferson became president, he kept Hamilton's bank and purchased the Louisiana Territory—in spite of his doubts about the constitutionality of either.

One of my few criticisms of the book would be Chernow's treatment of many of the Founding Fathers. (I doubt you are accustomed to reading about how petty, devious, and partisan these otherwise great men could be.) He is especially prone to Thomas Jefferson bashing and, in my opinion, doesn't come close to giving James Madison his due. However, at times his observations can be quite entertaining. I loved it when he described Hamilton and John Adams as "two brilliant and unstoppable windbags."

So, what's my take? We live in a Hamiltonian world, and we should embrace it. But that doesn't mean we can't personally live a more Jeffersonian life—enlightened, skeptical, inquisitive, and rational. After reading this book, I believe Alexander Hamilton would agree with that too.

Papa

*(Author's note) My granddaughter Katie, an adolescent thespian enamored with the musical *Hamilton*, insisted on borrowing my copy of Ron Chernow's biography of Alexander Hamilton. To me, this 800-plus-page tome seemed like a stretch for any fourteen-year-old reader, even a bold, precocious one. This book review was my attempt to gently acquaint her with the task ahead. Of course, she didn't need any of my words. The book was later returned to me—read.

Wisdom of the Founders

—and by the way in the New Code of Laws
which I suppose it will be necessary for
you to make I desire you would
Remember the Ladies and be more generous and
favourable to them than your ancestors.
—Abigail Adams to John Adams, March 31, 1776

Those interested in American history frequently come to admire Founding Fathers such as Thomas Jefferson and John Adams, and often discover that these sages' wisdom still resonates today.

Even though ten of the first twelve U.S. presidents were slave owners, many of us still find it difficult to admire the often-quoted Thomas Jefferson as fully as we might have had he not been one of those ten. Many reasons are put forward to explain or justify Jefferson's ownership of others. Jefferson was an eighteenth-century Virginia planter, born and raised in a slave-owning family and aristocracy. By owning slaves, Jefferson was emulating the behavior of his elders—those whom he admired and after whom he patterned himself. However, when he enrolled in the College of William and Mary, Jefferson came under the influence of William Small. Small—who became Jefferson's mentor—was a Scottish professor steeped in the Enlightenment. Under Small's tutelage, Jefferson may have come to appreciate widely held views about the immorality and injustice of slavery. Jefferson even made some honest efforts as a young politician to promote emancipation in Virginia, only to quickly realize that a man with those views could not have a political future in his Commonwealth and would meet only with the disdain and hostility of his Southern peers. In spite of this, Jefferson made a later attempt at emancipation on the national stage, when in 1784 he included an antislavery plank in the Territorial Governance Act he drafted for Congress. This antislavery plank—which would have outlawed slavery in all new states—failed to pass by one vote. A distraught Jefferson is said to have lamented, "Heaven was silent in that awful moment and now millions of unborn will be enslaved through no fault of their own for the lack of one vote in the Congress of the United States." Think of Jefferson's altered legacy on race—as well as subsequent U.S. history—had that provision passed.

Being a Virginia planter, with its obvious implications, is the traditional reason—really excuse—given to explain why the author of the Declaration of Independence could never give up his slaves. However, the sad truth may be simpler and more basic to who Jefferson was. To paraphrase humanities scholar Clay Jenkinson: "Without slavery, Jefferson could not have been Jefferson." In other words, Jefferson could not have been the multitude of things that he is celebrated for: scholar, author, architect, inventor, multilinguist, agronomist; the list goes on and on. Without slavery, Jefferson would not have owned a library that became The Library of Congress; he would not have been able to afford clocks or meteorological and scientific instruments, to collect fossils, or to invent a new kind of plow. While acknowledging that much of this was done on the backs of the hundreds of slaves he owned during his lifetime, it's still undeniable that Thomas Jefferson was a remarkable Renaissance man. During a 1962 White House dinner honoring Nobel Prize winners, President John Kennedy famously said, "I think this is the most extraordinary collection of talent, of human knowledge, that has ever been gathered together at the White House, with the exception of when Thomas Jefferson dined alone." Kennedy went on to tell the laureates that "Someone once said that Thomas Jefferson was a gentleman of 32 who could calculate an eclipse, survey an estate, tie an artery, plan an edifice, try a cause, break a horse, and dance the minuet." Without the plantations and slaves that he inherited, our third president would have had difficulty being the Jefferson of Kennedy's celebrated and humorous quote. Owning slaves allowed Jefferson to exercise his intellect and curiosity in ways that he was probably just too selfish—or self-absorbed—to relinquish. In what seems like the ultimate irony, partly because of the financial structure Virginia planters lived under—but largely because of his many extravagances—Jefferson died deeply in debt, in spite of his slaves and land.

Is there a flip-side to Jefferson—a Founding Father equally brilliant, patriotic, and dedicated to life, liberty, and the pursuit of happiness—but without this dark side? Enter John Adams. Picturing the short, blustering, and brutally honest Adams alongside the tall, gangly, reportedly shy and diffident Jefferson, it's hard for someone of my generation not to imagine these two icons as the "Mutt and Jeff" of early American politics. This irreverent simile, however, soon evaporates upon learning almost anything about either man. Jefferson and Adams were both voracious readers, steeped in history and the classics. The prose that each left us is admired even today for its clarity and beauty. Living and working together in Paris after the American Revolution, they became fast friends and came to admire one another tremendously. John's wife, Abigail—later a harsh Jefferson critic—also learned to love Jefferson.

John Adams was a New England Yankee, not a Virginia planter, and he seems to have lived the ideals that Jefferson wrote about so eloquently. In his book *Notes on Virginia*, Jefferson wrote, "Those who labor in the earth are the chosen people of God, if ever He had a chosen people. . . ." A working lawyer and farmer, Adams's own hands were in the earth—an activity that Jefferson obviously greatly admired, but usually delegated to the humans he owned. Always a man of modest means, John Adams's long service to his country was at the sacrifice of his own livelihood and financial security. At no time during his years of public service did Adams seek to enrich himself.

Interest in Adams has been increasing as contemporary historians—Joseph Ellis, David McCullough, Gordon Wood, and others—have explained him further to us. In *Friends Divided: John Adams and Thomas Jefferson*, Wood says, "Jefferson told the American people what they wanted to hear—how exceptional they were, Adams told them what they needed to hear—truths about themselves that were difficult to bear. Over the years Americans have tended to avoid Adams's message; they have much preferred to hear Jefferson's praise of their uniqueness." It seems that Adams may have always been known for straight talk. When he defended the British soldiers in the Boston

Massacre trials in 1770, he told the jury: "Facts are stubborn things: and whatever may be our wishes, our inclinations, or the dictates of our passion, they cannot alter the state of facts and evidence." In what became a bedrock of our Constitution, Adams declared, "that Power was never to be trusted without a Check." The more one studies John and Abigail Adams, the more one realizes that they, as much as anyone of their exceptional generation, epitomized the ideals of Revolutionary America.

While writing *John Adams*—David McCullough's second Pulitzer Prize-winning work—the author came to greatly admire his subject. In a more recent book *The American Spirit: Who We Are and What We Stand For*, McCullough published a collection of speeches that he's delivered at commencements, dedications, and other affairs of note. My favorite chapter in *The American Spirit* is "The First to Reside Here." This chapter is the speech McCullough gave in the year 2000 on the occasion of the 200th anniversary of the White House. In this speech, McCullough reminded his listeners that Adams was the first president of the United States to live in the newly-completed White House. Because Abigail was yet to join him, John spent his first night there alone. The following morning, the lonely Adams penned a letter to Abigail. McCullough says that Franklin Roosevelt "thought so highly of the letter, and of two sentences in it, that he had it carved into the wood mantelpiece in the State Dining Room. And when Harry Truman supervised the rebuilding of the White House, he insisted that the inscription remain where it is today." Later John Kennedy had the inscription carved into the mantelpiece in marble. These two sentences of elegant prose, perhaps as much as anything he wrote, help us appreciate the wisdom and ideals of our second president. "I pray heaven," Adams wrote, "to bestow the best of blessings on this house, and all that shall hereafter inhabit it. May none but honest and wise men rule under this roof."

This was written by a man who must have realized that he was not going to be elected president for a second term—a second term that he coveted and that he may well have won had he succumbed to popular opinion and gone to war with France, a war Adams knew could have been disastrous for our new republic. Adams, like Jefferson and all other great men, was not perfect. McCullough acknowledges that "Adams could be vain, irritable, short-tempered." Although Adams kept us out of an ill-advised war, he signed the unjust Alien and Sedition Acts. But Adams always gave his heart and soul to his new country. He made repeated, perilous trips to Europe during the Revolution, secured desperately needed loans to finance the war, and signed the Treaty of Paris. The country that he helped found (he assigned Jefferson to write the Declaration of Independence and had suggested that Washington be appointed com-mander-in-chief) eventually adopted a constitution heavily influenced by the Constitution of the Commonwealth of Massachusetts that Adams had drafted years earlier.

In that letter to Abigail—written on the morning of his first full day in the new White House—is not only John's heartfelt wish and eternal blessing for the future of his country, but also a benchmark for his successors: "none but the honest and wise." Even the eloquent Thomas Jefferson could not have left us better advice.

Sins of Our Fathers

He had his medals and peace papers
to prove that he was the white men's friend.
But the Great Plains was a world unto itself. Lean Bear
was just thirty feet from the soldiers when they opened fire.
The chief was dead before he hit the ground.
After the smoke cleared, several troops broke
ranks and pumped more bullets into his corpse.
As Lincoln had cautioned Lean Bear, his
children sometimes behaved badly.
—**Peter Cozzens, The Earth is Weeping**

The slave-owning patriot Thomas Jefferson is often accused of hypocrisy for writing that all men are created equal. The Father of Our Country, George Washington, fares better in this discourse because he freed his slaves— but only upon his death when he no longer needed them.

Judging our forefathers and ancestors is inevitable, and it can be historical quicksand.

Family history research now enjoys widespread popularity. This popularity has resulted in television offerings that uncover the family trees of modern celebrities; the end result usually being repeated exclamations of: "Oh my God!" I wasn't surprised when a prominent movie star, known for his liberal views and social activism, was obviously embarrassed—even ashamed—when he was informed that some of his ancestors were Southern slave owners.

I've looked into my family history enough to think it's unlikely that I descend from Dixie slave owners. I'm also sure that I have skeletons in my closet—we all do. Most of my ancestors were too poor to own another human being, and my Puritan forefathers did not rely on slave labor to scratch out a living in New England. They also didn't hesitate to exploit, or later massacre, natives whose land they coveted. If a virulent small pox epidemic (brought to the New World by Europeans) resulted in the decimation of a local tribe, these pious ancestors of mine gave thanks to God for providing them with such choice land.

Even the most fortunate genealogist knows only a fraction of his family history. If we could venture back far enough into prehistory, every Homo sapien's family tree would be littered with unsavory characters. The Creationists (those who believe in biblical inerrancy), have only to study the Old Testament to discover the murderers and rapists spawned by Adam and Eve. Many of us take a longer view of human history. For us, the opportunity to have ancestors to be ashamed of is much longer than the 6,000 years or so available to the Creationist. Geneticists tell us that all modern humans are descended from individuals that lived in Africa—so called "Mitochondrial Eve" and "Y-Chromosomal Adam"—at least 100,000 to 200,000 years ago. If you choose to include among your ancestors the primates that our genetic Adam and Eve evolved from, you can add millions of years and countless generations to your family tree.

Early primate and human history would not have been pretty. I doubt that these ancestors always asked nicely when choosing a mate, deciding on dinner, or inhabiting the cave of their choice. Many paleontologists now believe that Homo sapiens were responsible, in one way or another, for the disappearance of the Neanderthals (as well as other early human species) they encountered as they left Africa and spread across the globe. It's now genetically certain that there was interbreeding between our Homo sapien ancestors and the vanishing Neanderthals. Was it love or war? The sad truth is that you exist because your ancestors didn't lack the killer instinct. They were aggressive enough to survive—when others didn't.

Fast-forward thousands of years; what about the 12 U.S. presidents who owned slaves? This list includes Ulysses S. Grant, who freed his slave in 1859. Abraham Lincoln never owned slaves, but slaves worked in the White House during his presidency. Slavery was a huge sin that has stained our national consciousness. But does that make all slave owners sinners? We know that even some free African Americans owned slaves. Many Southern slave owners, including Robert E. Lee and Thomas Jefferson, wrote about the evils of slavery, wishing for an end to the institution. We can applaud Lee when he called slavery: "A moral and political evil in any century." However, when Lee added that he thought slavery was "a greater evil to the white than to the black race," his moral indignation seems flimsy and less admirable. (Like most white Americans of his time, Northerners included, Lee was quite paternalistic toward the slaves. He thought them in need of constant guidance from white Americans, and better off in a Southern cotton field than in Africa.) Besides supporting the economy of half the country, slave labor enabled Virginians—like Washington, Jefferson, and Madison—to devote the time and resources to country and politics that our fledgling nation needed. Perhaps we should find better ways to recognize and honor the generations of slaves who made these unwilling contributions. Doing so might enable us to more fully address this sordid and undeniable part of our national heritage.

The treatment of American Indians by European immigrants as they settled the New World is another source of national shame. Of course, this was nothing new. Near annihilation, even genocide, has been common for thousands of years all across the globe as more technologically "advanced" cultures displaced indigenous people. In some cases, those same indigenous people had been responsible for the demise of tribes and settlements standing in their way hundreds or even thousands of years earlier. Conflict of this type appears to be inevitable and results in acts of brutality and butchery on both sides. In the 1700s my fifth-great-grandfather, his brother, and two uncles, were killed by Native Americans. Who was the aggressor, and who was at fault, in that tragic encounter is lost to history. Not lost to history are the United States government's repeated attempts to dislocate, even destroy, the native people who resisted removal from their lands. Today, we condemn such treatment as a cruel violation of human rights.

When our ancestors' behavior conflicts with our own idea of what may be proper or moral, I find it easier to understand them by looking back through the lens of time. By taking into consideration the many limitations that were placed on our ancestors' knowledge, experience, and worldview, those of us in the twenty-first century can more fairly judge our predecessors. I hope that our descendents will be similarly lenient in their reflections on us.

The Laureates

Great minds have purposes;
others have wishes.
—Washington Irving

The Five Nobel Prizes—for Chemistry, Literature, Peace, Physics, and Physiology or Medicine—were established in 1895 by the provisions of Swedish philanthropist Alfred Nobel's will and first awarded in 1901. Because Nobel Prizes cannot be awarded posthumously, some of the brightest minds that ever existed are not represented on the long list of Nobel recipients. This has inspired me to consider who, among those who died before any Nobel Prizes were bestowed, would most have deserved this honor. Each of us, I imagine, would have our favorites, and I'm confident that, when forced to choose, each of us would invariably snub some eminent figures that others would consider slam dunks for these honors. This doesn't bother me in the least; there have been numerous actual laureates over the years who were genuine head-scratchers for many. I enjoy the music of Bob Dylan and recognize his particular genius, but upon hearing that he received the Nobel Prize for literature, I do admit to having uttered an involuntary "What?" Many also wondered out loud what President Barack Obama had done to merit the Nobel Peace Prize so early in his presidency.

So what criteria should I use to bestow these imaginary prizes? Because of huge gaps in my knowledge, my choices will largely be limited to those from Western civilization. However, I hope that my laureates, particularly in the sciences, are people who saw possibilities and patterns in nature and came to understand truths that their peers could not—and had not for centuries. Frequently, people of this caliber were simply ignored by the critics and experts of their time; sometimes, the only reward they received was an early or painful death. Many of these geniuses would be astonished, even bewildered, to learn that theirs are now household names, or that their works have been studied by subsequent generations. In his book *Sum: Forty Tales from the Afterlives*, neuroscientist and author David Eagleman imaginatively declares, "There are three deaths. The first is when the body ceases to function. The second is when the body is consigned to the grave. The third is that moment, sometime in the future, when your name is spoken for the last time." For me, Eagleman's fantasy is an intriguing way to ponder death. However, does his idea also imply that it may be an eternity before some of our laureates—authentic or imaginary—are finally allowed to rest in peace? Let's hope not.

Peace

It isn't enough to talk about peace,
one must believe in it.
And it isn't enough to believe in it.
One must work at it.
—Eleanor Roosevelt

*L*et's begin with the hardest first. Of the five awards, the Nobel Prize for Peace may garner the most attention each year. For a while, I wondered why I was having so much trouble comprising a list of candidates for my imaginary Peace Prize. Then, I realized that to study history is often to learn about one war—and one warrior—after another. Edward Gibbon, author of *The Decline and Fall of the Roman Empire*, is often quoted as saying, "History is, indeed, little more than the register of the crimes, follies, and misfortune of mankind." We adorn military leaders with dramatic names like "Mad Dog" Mattis, "Stonewall"

Jackson, and Alexander the Great. Who ever heard of Gandhi the Magnificent?

Speaking of Mahatma Gandhi, by stipulating that I would only consider those who died before 1901, I have eliminated the person whom I, at least, reflexively associate most with the word "peace." Born in 1869, Gandhi was assassinated in 1948 (age seventy-eight). Gandhi was nominated five times, the last nomination occurring shortly before his assassination. In 1948, the Nobel committee did not award the Peace Prize, stating that "there was no suitable living candidate" that year. In 2006, Secretary of the Norwegian Nobel Committee, Geir Lundestad, said, "The greatest omission in our 106-year history is undoubtedly that Mahatma Gandhi never received the Nobel Peace Prize. Gandhi could do without the Nobel Peace Prize, whether the Nobel committee can do without Gandhi is the question." I admit to not even recognizing the names of many of the actual recipients of the Peace Prize; of those I do know, only a few—perhaps Jane Addams, Albert Schweitzer, Martin Luther King, Jr., and Nelson Mandela—seemed anywhere near as deserving as Gandhi would have been of the honor. Nevertheless, Gandhi was never awarded the prize, and, unless I changed my rules, he couldn't be my choice either.

As I mentally cycled through historical figures whom I thought many might suggest for my prize, I realized how few potential candidates—and even actual winners—lived lives of real peace. President Theodore Roosevelt won the Nobel Peace Prize in 1906 for negotiating an end to the Russo-Japanese War. While Roosevelt is a giant in American history and a man to admire for many reasons, he could not objectively be labeled a peace-loving man. Always ashamed that his father had not served in the Civil War, but rather paid another to take his place, Theodore enthusiastically waded into the Spanish-American War in a uniform custom-made from Brooks Brothers. Roosevelt always considered the day he led his Rough Riders up the hill in battle to be "the great day of my life" and "my crowded hour." He loved to be referred to as "The Colonel."

From looking at the list of the actual winners (Menachem Begin, Nelson Mandela, Woodrow Wilson), it is obvious that the Nobel committees have not considered conducting a war, or leading a paramilitary or terrorist group, to be disqualifying—provided there was a notable later accomplishment that significantly promoted peace. But I want my laureate to be someone whose entire life could serve as an example for the rest of us.

I'm aware that many would choose Jesus, Muhammed, or the Buddha. However, due to the wars, inquisitions, and persecutions that followers of major religious figures have fomented in their founders' names, I looked elsewhere.

(Even Buddhist organizations in Japan may have supported the government's war effort in World War II.)

When considering peace candidates, my thoughts initially landed on Henry David Thoreau (1817–1862). Even though, according to Ralph Waldo Emerson, Thoreau "kept his haughty independence to the end," Henry was a gentle, peaceful man who probably felt more kinship with Mother Nature (and Mother Nature with him) than just about anyone I can imagine. He was also a committed abolitionist and very active in the Underground Railroad. According to Laura Dassow Wells in *Henry David Thoreau: A Life*, "Henry regularly escorted escaping slaves to the northbound train, bought them tickets, ensured they had money, and either boarded with them at Concord (sitting at a distance keeping guard) or drove them to the whistle-stop at West Fitchburg." Thoreau also opposed the Mexican-American War, knowing victory would mean the spread of slavery. (Probably the only nonpeaceful chord Thoreau ever struck was his vocal support of John Brown, whom Thoreau, as an abolitionist, regarded as a hero.)

Few today, it's safe to say, know that Thoreau was very interested in and respectful of Native American culture as well as East Indian spiritual thought and yoga. He was also an early advocate of a healthy diet. In a society full of intemperate drinking, Thoreau abstained, claiming, "I believe that water is the only drink for a wise man."

Thoreau is best remembered, however, for his writing. Far from a best-selling author in his time, he would be flabbergasted to discover that anyone in our era would even recognize his name. His most famous book is *Walden*, a contemplation on living intimately with nature. However, another work, his essay "Civil Disobedience," had a profound effect on peace activists such as Mahatma Gandhi and Martin Luther King, Jr. In his eulogy of Thoreau, Ralph Waldo Emerson was more prescient than he, no doubt, ever realized when he declared, "The country knows not yet, or in the least part, how great a son it has lost." Upon reconsideration, Thoreau may be an even more attractive candidate for my literature prize!

With Henry David Thoreau safely tucked away for later consideration, I am awarding my imaginary Nobel Peace Prize (drum roll . . .) to **Roger Williams** (1603–1683), really my preference all along. While probably not a household name, I hope that you, dear reader, will also find him deserving of this honor. Williams was a Puritan minister who was banished from Massachusetts Colony in 1635 for what James A. Warren in *God, War, and Providence* called "his trenchant criticisms of Puritan religious intolerance and its exploitive Indian policies." Williams was one of the few people of his time who, according to Warren, believed that the "English had no right to settle on Indian lands without their consent [and fair compensation] and that the

Indians possessed a culture that deserved respect rather than condemnation." Roger Williams admired the Native Americans and became their trusted friend. In fact, again according to Warren: "Their interpersonal ethics often seemed to him [Williams] superior to those of Christian Europeans." In particular, Williams noted that "scandalous sins" such as gluttony, robbery, murder, adultery, etc. were almost unheard of in the Narragansett people he observed. Williams was the founder of Providence Plantations (which became Rhode Island Colony) and maintained peace between the colony and the Native Americans for many years, even if it meant twice surrendering himself as a hostage. Williams was one of the few colonists in New England that the Narragansett people actually trusted.

Although the Puritans were willing to use Williams to mediate disputes and negotiate agreements with the Natives, the governments of Massachusetts, Connecticut, and Plymouth Colony—apparently distrustful of anyone cooperating with the Indians—formed an alliance against Rhode Island and the Narragansetts. This prompted Williams to sail back to England and secure a charter for his colony that ensured its safety. This charter also amazingly allowed both majority rule and religious freedom. Roger Williams had, against incredible odds, managed to found one of the most religiously tolerant governments in the world. Williams's government allowed the Quakers safe refuge at a time when they had been hanged in Boston. Rhode Island also became a safe haven for non-Puritans, lapsed Puritans, and Baptists, to name a few. In addition, Jews fleeing the Inquisition in Spain and Portugal were allowed to live in Newport, Rhode Island, creating in the mid-1600s the second-oldest Jewish congregation in America.

In addition to advocating justice and fair-play for the Native Americans, and founding a colony with previously unheard of religious freedom, Williams strongly opposed slavery, which was then legal in Massachusetts, Connecticut, and Plymouth colonies. Much to his chagrin, I'm sure, Williams was unable to ban slavery in Newport. In an unfortunate footnote to history, Rhode Island became even more complicit in the slave trade after Williams's death.

In *God, War, and Providence*, Warren summarizes very well the reasons that Roger Williams is deserving of our notice: "In establishing a colony in Rhode Island that respected Indian customs and rituals to a far greater extent than other English settlements, in challenging orthodox Puritan assumptions about the inseparability of Christianity and civility, and the church and state, Roger Williams was presenting a strikingly original, alternative version of what America should look like. In this respect, he was a man whose ideas were many years ahead of his time."

I bestow my Nobel Peace Prize on **Roger Williams** believing that if enough others of his time—and since—had emulated his life in both words and deeds, the moral arc of America would have proven decidedly more just.

Physics

*Galileo and Kepler had "dangerous thoughts,"
and so have the most intelligent
men of our own day.*
—Bertrand Russell, The Conquest of Happiness

hysics can be thought of as two almost distinct disciplines, one being so-called Newtonian physics, and the other being quantum physics. Because quantum theory was not really formulated and accepted until the early twentieth century, the winner for early physics might seem pretty obvious. In 1687, the English physicist, astronomer, and mathematician Sir Isaac Newton (1642–1726/27) published *Philosophiae Naturalis Principa Mathematica*, laying out for the benefit of humanity his laws of motion and universal gravitation. In *The Day the World Changed*, author James Burke reminds us of *Principa's* significance: "With the theory of universal gravity, Newton destroyed the medieval picture of the world as a structure moved by the unseen but ever-present hand of God. Man was no longer at the centre of a system created for his edification by the Almighty; the earth was merely a small planet in an incomprehensibly vast and inanimate universe which behaved according to laws that could be calculated." Certainly, his theory of gravity is what Newton is most famous for—and, no doubt, always will be—but he was also a pioneer in the fields of mathematics and optics. He was even the first to calculate the theoretical speed of sound.

Unlike many other geniuses of the past who came up with groundbreaking theories, Newton did achieve wide recognition and some fame in his own time. He held a named professorship of mathematics at the University of Cambridge, served briefly as a member of Parliament, and, of course, was knighted by Queen Anne in 1705. As a Fellow of the Royal Society (elected in 1672 and president from 1703 until his death), Sir Isaac was recognized by his peers as the foremost "natural philosopher" of his time. Founded in 1660, the Royal Society is the oldest scientific institution in the world, with a membership that is a veritable "who's who" of influential scientists and thinkers—Charles Darwin and Stephen Hawking were two later Fellows. Newton is not only buried in Westminster Abbey but is also commemorated with a monument near his tomb.

In a letter to Robert Hooke, Newton acknowledged that he achieved what he had by "standing on the shoulders of giants." One such figure was the German astronomer and mathematician Johannes Kepler (1571–1630). In fact, one of the ways that Newton demonstrated the validity of his laws of motion and gravitation was by employing them to prove Kepler's laws of planetary motion. Kepler's were not the only shoulders upon which Newton stood when he formulated the laws for which he is so famous. Since the actual Nobel Prize can be shared by up to three winners, we should explore some more brilliant shoulders.

Polish polymath Nicolaus Copernicus (1473–1543) was, among other things, a mathematician, astronomer, physician, and linguist. (He also had a doctorate in Canon law and formulated an economic theory.) In 1514, Copernicus found time to publish a manuscript he called *The Little Commentary*. This work questioned the prevailing belief that the Earth was the center of the Universe, replacing it with the Sun, which was orbited by a moving Earth. This theory was more fully developed in Copernicus's *On the Revolution of the Celestial Sphere*, published just after his death in 1543. James Burke, again in *The Day the World Changed*, says of this publication, "In the theory he proposed a system heliocentric in nature, with the earth orbiting the sun and spinning once a day on its axis." Burke also surmises, "The fully developed helio-centric argument was not published until Copernicus's death in 1543. This might be evidence of Copernicus's awareness of the effect his new theory would have, or it may be that he simply felt it would be misinterpreted." It appears that Copernicus went to his grave unscathed by his beliefs; however, later advocates of Copernicus's theory, participating in what is called the Copernican Revolution, did not fare so well.

Giordano Bruno (1548–1600) was an Italian Dominican Friar born soon after Copernicus's death. Bruno was a philosopher, poet, and mathematician, but he is best known for his view that the Universe was infinite and contained celestial planets like the Earth. He also believed that the stars were suns with their own planets, and that these planets might have life of their own! It's popularly believed that Bruno was burned at the stake by the Roman Inquisition for these beliefs, and is often held up as a martyr for science. In truth, these heretical cosmological beliefs were just a part of the indictment that sent him to his awful death. He also had numerous theological differences with Rome—always a dangerous stance to take in his era—and was condemned as a "magician" (apparently not the entertaining kind).

In spite of his bravery and foresight, Bruno seems, at most, a sideshow compared to another early scientific giant: Galileo Galilei (1564–1642). In *Galileo and the Science Deniers*, astrophysicist Mario Livio writes that Albert Einstein considered Galileo "the father of modern physics—indeed of modern science altogether." Einstein also considered "Galileo's discovery and use of the scientific method" to be "one of the most important achievements in the history of human thought." James Burke (again in *The Day the World Changed*) also helps us understand why Einstein may have felt the way he did about Galileo's work: "Galileo brought about an intellectual revolution by proposing that physicists should dispose of Aristotelian essences [an Earth-centered Cosmos, among other things]. His [Galileo's] view was that the only way to find out what was happening was to observe and experiment; that in experiment one should look for the nearest cause for a phenomenon, and for events or behavior that were regular in occurrence, which could be reliably observed by the senses; and that everything should be reduced, if possible, to mathematics."

How was Galileo able to make such a fundamental and profound impact on science and human thought? It may seem almost providential that a new technology—the telescope—entered the world scene at a time when Galileo could make such marvelous use of it. An enthusiastic Galileo—we would now call him an early adopter—built his own telescopes, which he continued to improve and even sold to others. Because observers learned to trust what they saw through telescopes on land and sea, they could not readily dismiss Galileo's drawings or his descriptions of the observations he made when he pointed his telescope to the heavens—like the satellites of Jupiter and the mountains on the moon (mountains that Aristotle had taught his students could not exist, as he believed that the moon and other heavenly bodies were perfect spheres composed of "aether" or "quintessence"). Peering through his telescope, Galileo became a vocal advocate of the Copernican view of the Cosmos.

The Catholic Church was more than a little disturbed by cosmological theories that blatantly contradicted the Church's teachings about the Universe (e.g., a stationary Earth at the center of the Universe) that were based on passages from the Bible. Galileo knew the pope personally, and he seems to have overestimated how fond the pope was of him—as well as the liberties that he himself could afford to take. In 1632, Galileo published *The Dialogue on the Few Chief Systems of the World*. Instead of writing an academic treatise promoting his theory, Galileo penned a spirited dialogue, no doubt thinking that a fictional debate would be more acceptable and perhaps be seen as less absolute. James Burke tells us what went wrong: "It [the *Dialogue*] caused a sensation. The book showed the opponents of the Copernican system to be simpletons which was seen to be a full-blooded attack on the church." In 1633, after a gut-wrenching trial, Galileo was condemned to house arrest in Arcetri, near Florence, where he remained until he died in 1642. Galileo was forced to disavow his own beliefs about the Cosmos—beliefs that he felt sure were valid. Before we judge him too harshly, bear in mind that it was only by recanting that Galileo—by then an aged man—escaped torture and/or a hideous death at the hands of the Inquisition (just recall Bruno).

There's an anecdote that has persisted for centuries, and although it may be every bit as apocryphal as George Washington's cherry tree, I, for one, want it to be fact: After denying what he knew to be true and exiting the trial, Galileo is said to have murmured of the Earth: "And yet it moves." Regardless of that little tale's veracity, Galileo left no doubt about how he felt when he wrote that he did not believe that "the same God who has given us our senses, reason, and intelligence wished us to abandon their use," and that "the Bible shows the way to go to heaven, not the way the heavens go."

Even though Galileo's book was removed from the Vatican's *Index of Prohibited Books* in 1835, it took much, much longer for the Catholic Church to acknowledge that Galileo was correct. In 1980, the Vatican announced it would undertake a new study of the entire Galileo matter but did not issue its final report until 1992. Livio tells us the pope then admitted the following: "The majority of theologians did not perceive the formal distinction that exists between the Holy Scripture in itself and its interpretation, and this led them unduly transferring to the field of religious doctrine an issue which actually belongs to scientific research." (Whew . . .) To some, this pronouncement may sound more like a feeble excuse than much of an apology, and the pope finished by reassuring his flock that "science and religion are in perfect harmony." There is no explanation as to why it took the Vatican centuries to admit its error, and Livio, in *Galileo and the Science Deniers*, tells us that the "media around the world had a feast," such as in the headline from the *New York Times*: "After 350 Years, Vatican Say Galileo Was Right: It Moves." The *Los Angeles Times* took a similar tactic: "It's Official: The Earth Revolves Around the Sun, Even for the Vatican."

Enough! Forced to choose from so many great minds in history, I award my fanciful Nobel Prize in Physics to be shared by three winners whom I consider the most deserving: **Nicolaus Copernicus** (1473–1543), **Galileo Galilei** (1564–1642), and **Sir Isaac Newton** (1642–1726/27). These giants of human thought scrutinized the heavens above; instead of seeing outlines of gods and other mythical characters, they detected distant bodies obeying laws that they could often confirm with equations. Their observations and conclusions dared to challenge what nearly everyone else believed to be unquestionable and true. Thanks to Copernicus, Galileo, and Newton, humans gained a radical new understanding of the Universe around them and were better able to appreciate their place in it.

Chemistry

—from the oak of the forest,
to the grass of the field,
every individual plant is
serviceable to mankind.
—Sir John Pringle

If you've spent much time in a chemistry lab, you may remember learning about Boyle's law, even if you can no longer cite it. (Briefly: the pressure of a gas increases as the volume of its container decreases.) Born in Ireland to aristocratic English parents, Robert Boyle (1627–1691) was educated at the ancient and prestigious Eaton College in England and in 1663 became a Fellow of the Royal Society. Boyle is regarded as one of the founders of modern chemistry, and when considering candidates for my imaginary Nobel Prize in chemistry, we should at least make this mention of him.

Antoine Lavoisier (1743–1794) was another giant in chemistry with a long list of accomplishments to his name, including naming oxygen and hydrogen. At age fifty, his career was cut short when his noble station caught up with him, and he was guillotined in the French Revolution. However, we are not done with this eminent scientist, who will play into our later discussion.

I haven't been in a high school or college science classroom for years, but I suspect most still have a periodic table of elements posted in a prominent place. In the table, known chemical elements are arranged to demonstrate their relationships based on their chemical properties, atomic number, etc.—starting with hydrogen (atomic number one) and ending with oganesson (number 118). This table is about as important as it gets, but it can be downright intimidating. Like me, there must be others who are more than happy to have taken their last exam requiring knowledge of the periodic table. Russian chemist Dmitri Mendeleev (1834–1907)—considered the Father of the Periodic Table—had, by 1869, formulated an early version of the table. The reader may notice that Mendeleev lived long enough to have been eligible for an early Nobel Prize, and his exclusion may rival that of Gandhi's. It's possible that Mendeleev's failure to win the Nobel Prize was due to the lengthy time interval between his important work and the time when prizes were first awarded in 1901, but he was well recognized and lauded by other scientific organizations including the Royal Society. Mendeleev was awarded the Davy Medal in 1882 and the Copley Medal in 1905. In 1892, he was elected a Foreign Member of the Royal Society. He has not been forgotten, either—there is a Mendeleev crater on the moon!

Polymath is a descriptor that could apply to so many of our candidates in the sciences, but English polymath Joseph Priestley (1733–1804) may be the head of the class. I first became acquainted with Priestley when reading Steven Johnson's enlightening biography of him: *The Invention of Air: A Story of Science, Faith, Revolution, and the Birth of America*. Johnson depicts Priestley at sixty-one years old: "He was among the most accomplished men of his generation, rivaled only by [Benjamin] Franklin in the diversity of his interests and influence. He had won the Copley Medal (the Nobel Prize of its day) for his experiments in various gases in his thirties, and published close to five hundred books and pamphlets on science, politics, and religion." (I would add a fun fact: while seeking a cure for scurvy, Priestley invented soda water.) Johnson adds, "Priestley's discovery did nothing to fight scurvy, but it did create a taste for carbonation that would ultimately conquer the planet."

Priestley was born in Yorkshire, England. Recognized as a precocious child, his family made sure he had the best schooling and tutoring available. However, by age sixteen Priestley had rebelled against the strict Calvinism of his early upbringing and enrolled at Daventry, an academy operated by Protestant dissenters. When finished with his schooling, Priestley was a true scholar, fluent in six languages. Of his early career Johnson tells us that Priestley "spent his twenties preaching to small, dissenting congregations in Needham and Nantwich, offending a few parishioners along the way with his maverick theories on the divinity of Jesus Christ." Johnson adds, "The supernatural distortions of the Apostle Paul was a favorite subject." Obviously not fulfilled by his heretical parish work, Priestley published *The Rudiments of English Grammar*. Johnson says that this was "one of the first attempts to systematically map the structure of the English language with the rigor that scholars had long applied to Latin and Greek."

Priestley also soon became interested in both chemistry and electricity, and he established a home lab for his experiments. In 1767, he published *The History and Present State of Electricity, with Original Experiments*. This book went through four English editions and was translated into French and German. Johnson says, "Copies circulated around the world: the Italian electrician Alessandro Volta read it; Franklin sent multiple copies back to the colonies. (By 1788 it was part of the standard natural philosophy curriculum at Yale.) The book would remain the principal text on electricity for nearly a hundred years."

By the time Priestley published the *History*, he was enjoying a fruitful friendship with Benjamin Franklin, who was then living in London. In fact, Franklin became not just Priestley's friend, but his mentor. For some time, I've wondered how Franklin's exploits with the kite became so widely known and revered. Johnson answered that question and more when he wrote that in the *History* Priestley had "created an iconic portrait of his mentor [Franklin], and planted him in the Enlightenment pantheon alongside Isaac Newton. Franklin with his kite remains the defining image of the practical scientific ingenuity of the American Founding Fathers. And we have Joseph Priestley to thank for it."

If you research Priestley, the achievement mentioned most prominently in almost any source will be his "discovery" of oxygen. (Wait a minute—what about Lavoisier?) Out of sheer curiosity, I stepped into the past and took a 1984 *World Book Encyclopedia* down from the shelf. The author of the entry on Priestley correctly captured the nuance of Priestley's "discovery" stating that he "shares the credit for the discovery of oxygen with Carl William Scheele of Sweden," and correctly identified Antoine Lavoisier as the scientist who named the gas. (Priestley had called it "dephlogisticated air.") What you don't glean from the

World Book is that while Scheele apparently isolated oxygen two years before Priestley did, Scheele—who called it "fire air"—did not publish his findings until 1777, long after Priestley had published his own results in Book IV of his *Experiments in Air*. The oxygen work that Priestley is so widely known for occurred during his experiments of 1774-75, when he produced a gas that burned brighter than in common air. In a fascinating and revealing aside, Johnson relates how, shortly after his oxygen experiments but before publication, Priestley left on a European tour. At a fateful dinner Priestley "gave a riveting account of his experiments to an audience of *philosophes*, among them Antoine Lavoisier, who would soon be Priestley's rival." There's little doubt that what Lavoisier learned when Priestley, as Johnson puts it, "sits down to dinner with the scientific intelligentsia of France and happily spills the beans about his exciting new experiment," would help Lavoisier "complete the chemical revolution that Priestley had initiated in his Leeds laboratory." When Priestley was asked later about the conversation he simply said, "I never make the least secret of anything that I observe."

As famous as he is for the 1773-75 experiments, it is Priestley's earlier laboratory work of 1771-72, for which he was awarded the Copley Medal, that I find the most compelling. We should note that Priestley was able to assemble a home lab allowing him to experiment with gases because of two fairly recent devices: the air pump and the pneumatic trough. With the use of these tools, inquisitive scientists like Priestley were able to capture, manipulate, and move gases between containers, as well as to create vacuums.

It was interesting to learn in Johnson's book that Priestley's seminal work was obviously influenced by his childhood pastime of capturing spiders and observing how long it took for them to perish in sealed jars. Hard as it is now to fathom, the mechanics underlying an animal's expiration given a finite quantity of air was a mystery at the time of Priestley's youth. As Johnson explains, "Did the creatures somehow exhaust the air they were breathing—in which case what was left in the jar? Or were they poisoning their environment with some invisible substance they released?" It was also known that a lit candle would flicker and die in the sealed chamber. Armed with this knowledge and his basic lab paraphernalia, Priestley began to explore the secrets of life and death on our planet.

Instead of the spider of his youth, Priestley began by isolating a mouse in a sealed container. Of course, the mouse quickly expired. Next, in a moment of genius, Priestley wondered how long a plant would survive in the vacuum compared to his animal subjects. He used a mint plant pulled from the ground. To Priestley's surprise, the mint not only survived, but it also continued to grow! To add to the mystery, a lit candle placed inside the jar with the plant surprisingly was not extinguished either. Then Priestley did what you might expect of Priestley. As Johnson explains, "A mouse placed inside the jar with the plant could survive happily for ten minutes, while a mouse placed in a plant-free jar in which another mouse had previously expired would begin to convulse in seconds. Somehow the plant was disabling whatever it was that snuffed out the candle and suffocated the mouse." Priestley repeated the experiment again and again using different plants and found that the effect was not restricted to mint. (Spinach, interestingly, was the most effective.)

During this time, Priestley was not only corresponding with Benjamin Franklin, but also received Franklin as a visitor at least twice at his home in Northern England. In a letter to Franklin, Priestley summarized some of his findings: "I have fully satisfied myself that air rendered noxious by breathing is restored by sprigs of mint growing in it. You will probably remember the flourishing state in which you saw one of my plants. I put a mouse [in] the air in which it was growing on the saturday after you went, which was seen days after it was put in, and it continued in it five minutes without shewing any sign of uneasiness, and it was taken out quite strong and vigorous, when a mouse died after being not two seconds in a part of the same original quantity of air, which had stood in the same exposure without a plant in it." Johnson astutely says of

Priestley's experiment: "There was a system lurking in the glass that was a microsystem of a vast system that had been evolving on Earth for two billion years." A letter from Franklin back to Priestley hints that Franklin was sensing the system here and starting to understand its implications. "That the vegetable creation should restore the air which is spoiled by the animal part of it, looks like a rational system and seems to be a piece with the rest." Franklin was half right when he speculated that "the strong thriving state of your mint in putrid air seems to shew that the air is mended by taking something from it, and not by adding to it." Franklin, a man ahead of his time, added, "I hope this will give some check to the rage of destroying trees that grow near houses, which has accompanied our late improvements in gardening, from an opinion of their being unwholesome. I am certain, from long observation, that there is nothing unhealthy in the air in woods." Just think; in this correspondence, we are privy to two of the finest minds of the Enlightenment as they first begin to fathom the now commonly-known cycle of gases critical to all life on Earth—the absorption of carbon dioxide and production of oxygen by plants. Interested scientists in the Royal Society soon became aware of Priestley's work and had no difficulty appreciating its significance. When Priestley was awarded the prestigious Copley Medal, Sir John Pringle, president of the Society, gave a speech that many could both hear and heed in our time, including this excerpt: "From these discoveries we are assured, that no vegetable grows in vain, but that from the oak of the forest to the grass of the field, every individual plant is serviceable to mankind: if not always distinguished by some private virtue, yet making a part of the whole which cleanses and purifies our atmosphere."

This could certainly be the end of our story, but we're not done with this remarkable man. Because of his religious views, his radical Whig politics, and his support for the French Revolution (the French had bestowed honorary citizenship upon him), Priestley had become, in Johnson's words, "the most hated man in Britain." In 1791, there were riots in Birmingham; Priestley's life was threatened, and his home was burned to the ground. He gradually became convinced that he had no future in his home country of England—it was time to emigrate.

Priestley's prior support in Britain of the American Revolution, his scientific renown, and his persecution in England, ensured that he would be welcomed in the newly constituted United States, where one of his sons, Joseph Jr., had already settled. In 1793, Priestley and his wife sailed for America, where there was even a public debate over which state was the most deserving of his presence! Although his old friend and mentor Benjamin Franklin had died in 1790, Priestley still decided to settle in Franklin's

Pennsylvania. At first, it was smooth sailing for the new immigrant. He was invited to tea in Philadelphia several times by President Washington and became friends with another Enlightenment figure—physician, writer, and political theorist Benjamin Rush. He was also warmly welcomed by Vice President John Adams, whom Priestley had met in 1786, when Adams was the American ambassador in London. However, life is seldom tranquil for a man like Priestley. John Adams was elected the second president of the United States, and with the rise of political parties, began a politically contentious time in our new republic. It didn't take long before Priestley became close friends with the newly elected vice president, Thomas Jefferson, who had admired Priestley and his exploits for years. An ongoing Anglo-French conflict was driving politicians as well as ordinary citizens to choose sides, and Priestley naturally gravitated toward Jefferson's pro-French Democratic-Republican Party. Letters soon surfaced revealing that Priestley had been critical of the Adams administration and Adams himself. These letters placed Priestley at risk of imprisonment under the Federalist party-backed Alien and Sedition Acts. Although Adams had signed the Acts, he was never an enthusiastic enforcer of this draconian legislation. Perhaps proving that he was not as thin-skinned as he is often portrayed, Adams personally intervened to prevent his old friend from being prosecuted. Jefferson's subsequent election to the presidency began a time of relative political calm, even for Priestley. In fact, he wrote to a friend: "[For] the first time in my life (and I shall soon enter my 70th year), I find myself in any degree of favor with the governor of the country in which I have lived, and I hope I shall die in the same pleasing situation." Priestley's late-in-life wish was granted in 1804, when he died peacefully and without torment. Calm, perhaps, but never idle; the year before his death, he had published four new volumes in his *General History of the Christian Church*. At life's end, he had also been corresponding with Jefferson about curriculum suggestions for Jefferson's latest brainchild, the planned University of Virginia.

Years later, to the benefit of subsequent generations of Americans, John Adams and Thomas Jefferson renewed their friendship with a long procession of letters that traveled back and forth between Quincy and Monticello. The frequency with which Joseph Priestley figures in this now-famous correspondence (more often than Washington, Franklin, or Madison) has been noted by historians. In some letters, Priestley was a flash point for their differences and grievances, but it's also apparent that both of these Founding Fathers had been profoundly affected by our combative chemist. Jefferson has written that Priestley's life was "one of the few precious to mankind," and he later wrote to Adams about how spiritually important Priestley had been for him. Jefferson was often criticized for being a skeptic (or worse) of many tenets of Christianity. Jefferson wrote to Adams, confessing that: "I have read [Priestley's] *Corruptions of Christianity and Early Opinions of Jesus* over and over again; and I rest on them . . . as the basis of my own faith. These writings have never been answered." Johnson maintains that Priestley's ideas demonstrated to Jefferson that "Christianity was not the problem; it was the usurped counterfeit versions that had evolved over the centuries that he could not subscribe to." Jefferson could now embrace Christianity—albeit his and Priestley's kind—in his old age. Despite the discord and falling-out between Adams and Priestley, Adams admitted to Jefferson: "This great, excellent, and extraordinary Man, whom I sincerely loved, esteemed, and respected, was really a Phenomenon: a Comet in the System, like Voltaire."

If you are a student of history for long, you are bound to encounter historical figures that you can't help but admire. Perhaps risking personal ruin, they made a courageous stand, led others to freedom, wrote a book that changed the world, or even sacrificed their life for some noble purpose. There are still others that you find yourself in awe of because you had no idea that such an accomplished and improbable human being was even possible, a person like **Joseph Priestley** (1733–1804), the recipient of my Nobel Prize in Chemistry.

Medicine or Physiology

In common with most heart specialists,
I am a Hamlet who is always wondering what to do.
Fortunately, the patient improves or
is even cured during the soliloquy.

—George Sheehan, **Running and Being**

$\mathcal{N}$umerous physicians and scientists over the last 2,000-plus years have been crucial to the advancement of scientific medicine. A name familiar to many is **Hippocrates**, a Greek physician who lived from about 460 BCE to 370 BCE. I learned much about Hippocrates and the history of medicine from lectures in the Great Courses offering "Doctors: The History of Scientific Medicine Revealed Through Biography." (This course was presented by the late Yale surgeon and author Sherwin B. Nuland, M.D.) From Nuland, I learned that prior to Hippocrates, healing was largely superstition rooted in the supernatural. Hippocratic teaching was revolutionary: diseases are a part of life; the body can heal itself (thus "first do no harm"); listen to and examine patients, including their bodily discharges and vital signs; keep records so you can learn from your patients and make a prognosis; teach the next generation of healers what you have learned; and practice and live by a code. It may be impossible to know whether we should give credit only to Hippocrates or rather to the physicians who gathered around him for all these early teachings.

Galen of Pergamum (129 CE–c.200 CE) was a later Greek physician who lived in the Roman Empire and served as the personal physician to Emperor Marcus Aurelius. Because Roman law forbade dissecting humans, Galen was able to peer inside the human body only when caring for trauma victims such as injured gladiators. However, Galen was interested in how the body actually functioned, and in an effort to learn more, he dissected primates and other animals. He demonstrated numerous findings—for example, that urine is produced in the kidneys, not the bladder—and he produced paralysis and loss of feeling by cutting the spinal cord at various levels. As important as he was, Galen was wrong about much, and it took centuries for many of Galen's theories to be rejected. I was surprised to learn from Dr. Nuland that prescriptions created by Galen from botanicals, oils, and minerals—called Galenicals—were used until the early twentieth century.

In spite of the intervening centuries, one of the first to debunk Galen was Flemish physician and anatomist **Andreas Vesalius** (1514–1564). By the time Vesalius was a student at the University of Padua in Italy, human cadavers were being dissected, and Vesalius could see Galen's errors. However, when Vesalius attempted to expose Galen's inaccuracies, he was viciously attacked. In his book *The Great Influenza: The Epic Story of the Deadliest Plague in History* (which includes an engrossing history of medicine), author John Barry reminds us that Vesalius was not only verbally accosted for his efforts, but also sentenced to death. (Fortunately, the sentence was commuted.) However, Vesalius is important for more than his outspoken courage. He teamed with the talented artist Jan van Calcar, a protégée of the great artist Titian, and, according to Nuland, "left us the first accurate anatomy text ever written. He also demonstrated the importance of skepticism: the idea that nothing shall be believed that cannot be personally verified."

Many of the most elemental things about our anatomy and physiology were still misunderstood when the Englishman **William Harvey** (1578–1657) followed Vesalius as a student at Padua. For example, it was thought that food was transported from the stomach to the liver where it was transformed into blood (but no explanation was given for how the blood returned to the liver). After graduating from Padua, Harvey became one of the busiest and most prestigious physicians in London. With kings and aristocrats for patients, Harvey still found time to pursue what interested him most—how the heart functioned. Except for understanding how blood got from the most peripheral arteries to the most peripheral veins, Harvey figured it out. (Thirty-two years later, the Italian micros-copist Marcello Malpighi would solve that mystery when he demonstrated the existence of capillaries.) Of Harvey's singular accomplishment, Nuland says, "By general agree-ment, the greatest gift ever made by one man to the science and art of medicine was the discovery of the circulation of the blood by William Harvey." Barry seems to concur with Nuland and writes that Harvey's accomplishment was "arguably perhaps the single greatest achievement of medicine—and certainly the greatest achievement until the late 1800s." Harvey's all-important discovery, however, was not appreciated by his fellow healers. Nuland says, "Doctors continued bleeding, purging, and so on. They thought these methods worked and saw no reason to change . . . They found no practical use for the notion of the circulation of blood." It seems to me that doctors in this era may have been indifferent to Harvey's findings because they lacked later medical advances that made understanding blood circulation so critical. Even in 1806, as we shall see, the scientifically well-informed Thomas Jefferson considered

Harvey's discovery, "a beautiful addition to the animal economy," but he couldn't fully appreciate its significance. William Harvey was simply ahead of his time.

While both Vesalius's and Harvey's accomplishments were not valued by their peers, this was not the case with the Italian anatomist and physician **Giovanni Battista Morgagni** (1682–1771). According to Nuland, Morgagni was "one of the most renowned physicians in the world. His greatest memorial is his book, which was one of the real turning points in the long history of medicine." What did Morgagni do to merit such accolades? During his long career—most of it spent as an anatomy professor at Padua—Morgagni dissected countless cadavers. Fortunately, he maintained a detailed record of the patients' illnesses as well as his findings at autopsy. It was not until he was 78 or 79 that Morgagni published the book for which he is so famous. In it, he included 700 case reports of patients that he had dissected. Some of the cases can be attributed to another famous anatomist under whom Morgagni had studied: Antonio Valsalva (himself a student of Malpighi). On publication of the book in 1761, Nuland says, "Doctors realized they should no longer look for the origin of the disease in the entire body, but in one specific place—a concept that came to be called the anatomic concept of disease: A doctor had to know anatomy in order to make a diagnosis . . . this is how the modern physical exam developed." With Morgagni, we may see the first real glimmer of modern medicine.

Thus far we've made little mention of surgery or surgeons. Hippocratic healers, as well as others, did perform some procedures such as fracture stabilization and skull trephination, but throughout medieval times and beyond, the surgeon/barber amputated limbs, lanced boils, etc., at the direction of the physician. According to Nuland, **John Hunter** (1728–1793), "brought science into surgery." (Of course, until anesthesia and antisepsis arrived, surgery could not enjoy the universal acceptance it's now accorded.) Born in Scotland, Hunter was an unenthusiastic student who dropped out of Oxford. He later moved to London where his surgeon brother, William, had opened an anatomy school. John proved to be so skilled at dissection that he was soon teaching at his brother's school. John later became a renowned teacher and surgeon and went on to research topics never before examined. He wrote the first book on inflammation, performed the first recorded case of artificial insemination, invented many of the terms that dentists use today (for example, *bicuspid* and *incisor*) and studied the healing process.

John Hunter was the kind of character you'd have trouble making up, and I've been fascinated with him since reading about him years ago in John Knobler's book *The Reluctant Surgeon*. Apparently, Hunter's true interest was comparative anatomy; in addition to human cadavers, he dissected a wide spectrum of animals. In order to display the numerous specimens that he collected—human and otherwise—Hunter turned his home into a museum, and he wasn't above body snatching or bribery to obtain what he wanted.

In Hunter's time, surgeons needed to "get in and out" quickly to avoid, or at least minimize, the catastrophic bleeding and excruciating pain that caused the demise of so many surgical patients. Hunter was such a superb technician that, in spite of his unusual extracurricular interests, he was named Surgeon Extraordinary to the King. According to Nuland: "In 1788 when Sir Percival Pott died, Hunter was acknowledged as the greatest surgeon in London." The tablet on his gravesite in Westminster Abbey reads, "The Royal College of Surgeons has placed this tablet on the grave of Hunter to record admiration of his genius as a gifted interpreter of the divine power and wisdom at work in the laws of organic life, and its grateful veneration for his services to mankind as the founder of scientific surgery."

It's likely you've seen a picture or dramatization of an early physician placing his ear on the chest of a patient. Supposedly, **René Laennec** (1781–1826), an especially shy young French doctor, was making his hospital rounds when, according to Nuland, "Laennec came to a new patient—a pretty, buxom young woman, who was said to have heart disease. He found himself unable to engage in the customary practice of placing his ear directly over the patient's chest." It's said that after thinking this over, Laennec came back, rolled up a notebook, and placed an end on the woman's chest forming a cylinder through which he listened to her heart. Laennec was inspired to experiment further with a piece of tubular wood that he called a stethoscope. Laennec was not only clever but must also have been a skilled and knowledgeable clinician; he used his new invention to make careful observations that he published in 1819. This book not only showcased his new stethoscope, but it also presented a new method of classifying heart and chest diseases. Laennec became a professor of some renown before dying at age forty-one of tuberculosis. He considered the stethoscope "the best part of my legacy."

In the centuries before the emergence of scientific medicine, legions of suffering people would have sought out healers. Unfortunately, the remedies these healers had to offer might often have been worse than the patient's disease. However, one essential treatment that physicians had been armed with for centuries was opium. Laudanum, a tincture of opium mixed with a variety of other ingredients (one of which invariably was alcohol), was a popular form of opium. In addition to being dispensed by the physician, laudanum could be purchased on the open market. It was employed as a treatment for almost any ailment: pain, dyspnea, insomnia, cough, respiratory disease, diarrhea, nervousness; the list would be endless. It's been said (rightly, I think) that "the greatest pleasure in life is the relief of pain," and there must have been countless people who became dependent on laudanum. Laudanum, along with copious alcohol, was also commonly used preoperatively. I have no doubt that laudanum helped relieve headaches, diarrhea, and numerous other ailments. However, even supplemented with alcohol, laudanum would not have provided the relief patients needed when being cut to remove a bladder stone or trying to endure having their arm or leg sawed off! We can all be grateful for the advent of general anesthesia, but whom should we thank?

It turns out that our chemistry laureate Joseph Priestley discovered nitrous oxide in 1772, and it didn't take long before others discovered that inhaling or sniffing it caused a type of euphoria. People would pay traveling "chemists" to attend their "laughing gas parties." Ether had been synthesized even before Priestley discovered nitrous oxide, and in the nineteenth century, sniffing ether also became a recreational pastime. The person to first demonstrate the successful use of general anesthesia, at least in a documented setting, was William Morton (1819–1868), a New England dentist. Morton had anesthetized animals with ether and asked John Collins Warren, the senior surgeon at Massachusetts General Hospital, if he could anesthetize one of Warren's patients. Warren allowed Morton to anesthetize a patient with a challenging jaw tumor whose removal would take longer than Warren thought the patient could tolerate. On October 16, 1846, in front of a gallery of medical students and staff, Morton anesthetized the patient. Nuland tells us that Dr. Morton then announced to Dr. Warren, "Sir, your patient is ready." Warren spent 25 minutes carefully dissecting and removing the tumor, and "the patient didn't move a muscle." Completing the operation, Warren announced, "Gentlemen, this is no humbug." Nuland says, "With these words the era of modern surgery began."

Does Morton deserve the credit—and the Nobel—for introducing general anesthesia? After what you just read, it might seem so, but the story of anesthesia's emergence borders on the bizarre. Morton, it turns out, had been a student of Dr. Horace Wells, a Hartford, Connecticut, dentist who—after sniffing nitrous oxide at a traveling chemist's performance—started using it on his patients to extract teeth. Before Morton's successful demonstration, Wells had given nitrous oxide to a student at Massachusetts General Hospital for a dental extraction. However, because the patient screamed during the extraction, those attending the experiment judged it to be a failure, even though the patient later said he had not felt any pain. Morton, who realized the importance of his successful demonstration, envisioned obtaining a patent on ether and wouldn't reveal the identity of the anesthetic substance. Charles Jackson, a Harvard chemistry professor, then claimed he had suggested to Morton that Morton use ether as an anesthetic—and Jackson wanted his share of the credit. Morton and Jackson together obtained a patent in which the ether was called "letheon." Now Wells, the dentist who had been experimenting with nitrous oxide, claimed that he was responsible for the entire idea of anesthesia. Wells, however, was becoming mentally imbalanced and was arrested after pouring acid on a prostitute in the street. Wells smuggled chloroform and a knife into his jail cell and killed himself by opening his femoral artery. Crawford Long, a small-town Georgia doctor, then came forward claiming that he had used ether on his patients four years before Morton's famous demonstration. In a rather peculiar attempt to resolve this miasma, Congress intervened, offering a prize of $100,000 to the individual who could best prove that he deserved the credit. Morton and Jackson infringed on their own patent by using ether freely during the Mexican-American War and were soon suing one another. Morton was then censured by the American Medical Association, suffered a stroke, and died. After viewing Morton's gravestone declaring that Morton had discovered anesthesia, Jackson became psychotic and died in a mental hospital. Crawford Long decided not to press his claim and died peacefully in 1878. Congress never awarded the $100,000 prize.

A reader already familiar with some of the important names in the history of medicine might expect to encounter **Robert Koch** (1843–1910) in this discussion. Koch, a German physician, is one of the founders of modern bacteriology. He has been well recognized for identifying the microorganisms causing tuberculosis, cholera, and anthrax. Koch was one of the early pioneers in scientific medicine who was fortunate to live long enough to be awarded the actual Nobel Prize in Medicine (in 1905, four years before he died). Another physician we can all be grateful to—**Joseph Lister** (1827–1912)—also survived into the Nobel era, but he never received the prize. Lister, however, accumulated many other honors in his lifetime. He is said to have been an object of worship by some in his native England, as well as in Germany and France; his eightieth birthday was celebrated around the world. By the time Queen Victoria and her successor Edward VII were finished honoring him, Lister could be addressed as either "Sir" or "Lord." Lister served as president of the Royal Society for five years and has continued to be honored long after his death. There is a Lister monument in London, a Lister statue in Glasgow, a Lister Institute, a Lister Hospital, and a Lister Medal. What Lister did to be so exalted was to take Louis Pasteur's work—determining that bacteria were the cause of putrefaction—and apply it to medicine and surgery. Before the adoption of Lister's antisepsis technique, surgery—including some relatively minor procedures—was accompanied by an approximately 45 percent mortality rate, usually from infection. It's not difficult to understand why Lister has been the recipient of so much admiration, even worship.

Before the advent of agriculture and the domestication of animals (approximately 12,000 years ago) there may not have been sufficient numbers of people living in close enough proximity to one another to sustain a viral or bacterial epidemic. Hunter-gatherers also did not live near—or with—their food source: animals whose diseases would in some cases cross over species and infect humans (as with influenza, for example). For these reasons, it's thought that epidemics were a product of the Neolithic era when herding and farming began. Since then, certain epidemic diseases like plague have been scourges of humanity, leaving huge imprints on civilization.

Smallpox has also been a sorely feared killer. The earliest evidence of smallpox is in 3,000-year-old Egyptian mummies, but there is no reason not to suspect that it's been with us much longer. At any time in the last few thousand years, many of us would have had ancestors who suffered or died from smallpox. In his comprehensive review of smallpox for the *Baylor University Medical Center Proceedings*, physician Stefan Riedel quotes case-fatality rates of 20 to 60 percent in some smallpox epidemics and even worse in others. In eighteenth-century London, for example, the mortality among infants was nearly 80 percent. Riedel says that during a late-1800s epidemic in Berlin, the case-mortality rate for infants was 98 percent.

As awful as it was in the Old World, smallpox in the New World was utterly catastrophic. Many historians now believe there was a mass die-off affecting the indigenous natives a few years before the arrival of the Pilgrims. In 1617, not long before the *Mayflower* landed, a pandemic of some type of infectious disease had swept through New England. Similar pandemics had also affected the natives in other parts of North and South America. French and English fishermen had fished off the Massachusetts coast for many years. They had contact with the local natives when they landed to collect firewood and fresh water. They were also known to capture Native Americans and sell them into slavery. These fishermen, and other early explorers and trappers, may have transmitted infectious diseases such as smallpox, influenza, chickenpox, plague, and hepatitis to the indigenous population. Unfortunately, the Native Americans' immune systems were totally naïve to European diseases, resulting in up to 90 percent mortality rates in some regions. Pilgrim Edward Winslow wrote of an area he visited: "Thousands of men have lived there which died in a great plague not long since; and pity it was and is to see, so many goodly fields, and so well seated, without men to dress and manure the same." In 1622, another colonist noted, "In this by where we live, in former time hath lived about two thousand Indians."

Of the dreadful diseases that infected the indigenous population of the New World, smallpox must have been one of the deadliest—and most grotesque. The pilgrim William Bradford's writings are now commonly quoted and remain a valuable source of information about early colonial New England. Among his recollections is a chilling account of a smallpox epidemic that he witnessed among a local tribe: "A sorer disease cannot befall [the natives], they fear it more than the plague. For usually they that have their disease have them in abundance, and for want of bedding or linen and other helps they fall into a lamentable condition as they live on their hard mats, the pox breaking and mattering and running one into another, their skin cleaving by reason there of the mats they lie on. When they turn them, a whole side will flay off at once as it were, and they will be all of a gore blood, most fearful to behold. And then being very sore, what with cold and other distempers, they die like rotten sheep."

Knowing that survivors of smallpox became immune to the disease, the practice of inoculation (or also called variolation) was, according to Riedel, "likely practiced in Africa, India, and China long before the 18th century, when it was introduced into Europe." Inoculation involved the instillation of "fresh matter from a ripe pustule" from a smallpox sufferer under the skin (subcutaneously) of a nonimmune person. The hope of inoculation was that it would induce a less severe case of smallpox than would a case of smallpox acquired naturally—but still result in immunity. While it was far lower than the mortality rate for naturally acquired smallpox, inoculated patients still experienced a mortality rate of up to two percent, and the number of people willing to undergo this procedure was limited. Thomas Jefferson transported his family from Virginia to Philadelphia so they could be inoculated, and George Washington had his troops inoculated en masse, preferring to schedule their illness rather than having his ranks unpredictably decimated on the eve of a battle. (During the Revolutionary War, a nationwide smallpox epidemic killed four times as many Americans as did the British Forces.)

Except perhaps for general anesthesia's sad cast, any of the scientists discussed so far would be deserving of a Nobel Prize. Acknowledging that, I'm stepping outside of this august group to embrace someone that even Dr. Nuland did not include in his course. Considering my homily on smallpox, some readers may have guessed where I'm headed.

The way it's sometimes portrayed can sound rather quaint: small town general practitioner overhears a comment about smallpox and tries a bold experiment that changes history. While not untrue, there's more to the story, and the abbreviated version sells our hero short. The English physician and surgeon, **Edward Jenner** (1749–1823), had always been interested in science and nature. Orphaned at five, he was apprenticed at the age of thirteen to a surgeon/apothecary near Bristol. When Jenner was twenty-one years old, he moved to London to study at St. George's Hospital under Dr. John Hunter, who encouraged Jenner's innate interests in the natural sciences. After two years at St. George's, Jenner returned to practice medicine in his native Gloucestershire, where Riedel says Jenner was considered "capable, skillful, and popular." Besides doctoring, Jenner found time to author medical papers, play the violin in a musical club, write light verse and poetry, carry out experiments with human blood, and even build and launch his own helium balloon. He continued his study of nature, too, publishing a study of the cuckoo that earned him election to the Royal Society.

Even during Jenner's apprenticeship, he would have, as Riedel put it, "heard the tales that dairymaids were protected from smallpox naturally after having suffered from cowpox." Knowing that it was a relatively benign disorder, Jenner decided to determine if cowpox could be transmitted from one person to another. In 1796, Jenner took matter from the fresh cowpox lesion on the hand of a young dairy maid and inoculated an eight-year-old boy. After the boy recovered from a brief and mild illness, Jenner then inoculated him with matter from a fresh smallpox pustule. When he did not develop smallpox, Jenner concluded that the boy, his first test case, was immune. For some reason, Jenner's first paper describing his success was rejected by the Royal Society.

It's said that there were others before Jenner, even a Dorset farmer named Benjamin Jesty, who experimented successfully with cowpox to prevent smallpox, but here is where Jenner stands out: Jenner did not give up. He persisted in performing what he came to call "vaccinations" and privately published a booklet explaining the process. Jenner sent the vaccine wherever he could, and in spite of ridicule from some quarters, wouldn't abandon a procedure he knew would be life-saving. Even though these efforts adversely affected his private practice, Jenner never attempted to enrich himself from his vaccination work. Riedel sums up Jenner's importance nicely, saying, "It was his relentless promotion and devoted research of vaccination that changed the way medicine was practiced." Today smallpox has been eradicated worldwide.

In 1806, Edward Jenner received a letter from the President of the United States, who wrote, "I avail myself of this occasion of rendering you a portion of the tribute of gratitude due to you from the whole human family. Medicine has never before produced any single improvement of such utility. Harvey's discovery of the circulation of the blood was a beautiful addition to our knowledge of the animal economy, but on a review of the practice of medicine before and since that epoch, I do not see any great amelioration which has been derived from that discovery. You have erased from the calendar of human afflictions one of its greatest. Yours is the comfortable reflection that mankind can never forget that you have lived. Future nations will know by history only that the loathsome small-pox has existed and by you has been extirpated."

President Thomas Jefferson's eloquent words of gratitude are a fitting tribute to **Dr. Edward Jenner** (1749–1823), the winner of my Nobel Prize in Medicine or Physiology.

Literature

Books are a sort of cultural DNA,
the code for who, as a society we are,
and what we know.
—Susan Orlean, The Library Book

*I*n his book *Everything in Its Place: First Loves and Last Tales*, the author and neurologist Oliver Sacks wrote about participating in a panel discussion titled "Information and Communication in the Twenty-First Century." One of the participants, an internet pioneer, proudly related how his daughter "surfed the internet twelve hours a day and had access to a breadth and range of information that no one of a previous generation could have imagined." Upon considering this, Sacks asked if she had read a Jane Austen novel, or any other classic novel. He was informed by the panelist that his daughter didn't "have time for anything like that." Sacks then wrote that he "wondered aloud whether she would then have no solid understanding of human nature or society, and suggested that while she might be stocked with wide-ranging information, that was different from knowledge; she would have a mind both shallow and centerless." He reported that "Half the audience cheered; the other half booed." Sacks may have been a bit hard on the other panelist, but his Socratic-like question reminded me just why it is important to award a Nobel Prize for Literature.

I spent my undergraduate college years as a premed major. While it might be unfair to call my long-ago pre-medical curriculum an anti-English major, I can't help but remember it as a non-English one—barely a whiff of literature or writing. Even now, after a lifetime of reading, I'm no literary savant. The most that I can hope for, I suppose, is to be a bit like Supreme Court Justice Potter Stewart, who in 1984 admitted that he had difficulty defining obscenity, but declared, "I know it when I see it." I still struggle when trying to explain why a particular book is great, but I'm hopeful that "I know it when I read it."

There were a limited number of deserving chemists, physicists, and physicians who lived before the Nobel era, but, my goodness, during that same time in history there were so many tremendous authors and great works! How do I even begin to choose who will win my imaginary Nobel Prize? Do I consider an author's body of work, a book's or author's influence on ensuing civilization, or the author's innovation, or do I simply award the prize to the author of the "greatest book of all time?" Good luck with that! Should the winner be someone who is still avidly read today (e.g., Austen, Dickens), or should I also consider those authors who are now read mostly when assigned (Melville comes to mind)? The list of candidates, in no particular order, is beyond impressive: Dante, Dostoevsky, Homer, Shakespeare, Chaucer, Dickens, Melville, Irving, Thoreau, Cervantes, Wordsworth, Whitman, Coleridge, Dickinson, Eliot, Shelly, Byron, Flaubert, Hawthorne, Charlotte Bronte, Emily Bronte, Poe, Swift, Hugo, Emerson, Voltaire, Moliere, Austen, Stevenson, Thackery, Defoe, Trollope, and many others. If you're wondering why they're not on the list, Tolstoy, Twain, Kafka, Chekhov, and Hardy all lived too long into the twentieth century to be eligible for my prize.

There is probably no work that has exerted a greater influence on subsequent literature and on Western civilization than the Bible, and some of its passages—particularly in the King James version—are among the most beautiful in the English language. However, because the Bible has multiple authors, many of whom remain unknown, determining to whom the prize should be awarded would be problematic.

A writer unlikely to appear on anyone's list of "the greatest," Michel de Montaigne (1533–1592), nevertheless deserves our mention as he singlehandedly concocted the memoir. As author Sarah Beckwell put it in her book on Montaigne: "This idea—writing about oneself to create a mirror in which other people recognize their own humanity—has not existed forever." Today, the memoir is a staple of literature and comes in all flavors. (A memoir ordinarily addresses only a certain chunk or fragment of the author's life; an autobiography would be more comprehensive, encompassing the writer's entire life to that point.) Late in his life and dying of cancer, President Ulysses S. Grant composed his memoirs—considered one of the best of this genre ever written. I'm sure this work profited from Mark Twain's editing, but Grant may have been a natural. I did not realize that a commanding general could come so close to harm's way until he told about how a horse was shot out from under him. The following passage demonstrates his vivid, graphic, and honest style: "One cannon-ball passed through our ranks, not far from me. It took off the head of an enlisted man, and the under jaw of Captain Page of my regiment, while the splinters from the musket of the killed soldier, and his brain and gore, knocked down two of the others." Grant was not afraid to address the horrendous effects of the battles in which he sent men to die, either: "The woods were set on fire by the bursting shells, and the conflagration raged. The wounded who had not strength to move themselves were either suffocated or burned to death." Impoverished at the end of his life as a result of bad investments, Grant's *Memoirs* sold well enough to provide for his wife and family after his death.

In the Peace section we considered Henry David Thoreau, and because of the far-reaching influence of his writing, I foresaw him as a candidate in the literature category. Now, looking over my long list of celebrated prospects, I'm afraid the competition may be too great here too. Why do I keep considering Thoreau? Perhaps it's tough to dismiss someone who over 150 years ago wrote, "It's not what you look at that matters, it's what you see." Thoreau also intrigues me not only because he was such an unexpected character (odd, perhaps?), but also because of the love-hate relationship some still have with him today. It's true that he lived for a time by himself in the wilderness, but it's also a fact that his cabin was close enough to home that his mother could stop by and pick up his laundry. In her biography of Thoreau, Laura Dassow Walls says, "His two years, two months, and two days living at Walden Pond became and would forever remain an iconic work of performance art." I would add that his overnight stay in the local Concord jail for refusing a pay a poll tax—from which sprang the famous "Civil Disobedience" essay—was so benign and so unlike Nelson Mandela's incarceration as to be laughable. In his book, *Fantasyland: How America Went Haywire: A 500-year History*, Kurt Andersen claims that Thoreau "invented a certain kind of entitled, upper-middle-class extended adolescence." Whatever you make of him, Thoreau was surely one of a kind. He self-published his first book, which meant he paid for the printing. When it sold poorly, he was forced to bring the unsold copies home. This prompted him to tell someone that "I now have a library of nearly nine hundred volumes, over seven hundred of which I wrote myself." Thoreau was a religious skeptic, and I've read in more than one source that his orthodox Aunt Louisa visited him on his deathbed and inquired, "Henry, have you made your peace with God?" Thoreau is said to have calmly replied, "I did not know that we had ever quarreled, Aunt." I don't know—maybe he was just a wise-guy, but he's also the guy who wrote, "The mass of men lead lives of quiet desperation. . . ."

But—if I were to award my literature prize to one individual on the basis of the influence of their written work, my choice would again be someone other than Thoreau. When he published *On the Origin of Species by Natural Selection* in 1859, Charles Darwin rocked the scientific world, and the shock waves are still being felt today. *Origin,* along with *The Descent of Man* published in 1871, transformed the way humans regarded themselves, their beginnings, and their place in the animal kingdom. Given the meager fossil record available to him and the contemporary lack of understanding of genetics and the other life sciences, the accuracy of his theory and of his predictions is astounding. Not only was he right about big things—like Africa being the site of human origins—but he aced the smaller stuff too. In *River of Consciousness*, Oliver Sacks gives us one such small example: "Examining one Madagascan orchid with a nectary nearly a foot long, he [Darwin] predicted that a moth would be found with a proboscis long enough to probe its depths; decades after his death, such a moth was finally discovered." Well before his seminal work on natural selection, Darwin was considered an expert on orchids, earthworms, beetles—you name it in the natural world. His reputation as a leading naturalist had been secured as a young man when he published a popular book about his now-famous voyage on the *Beagle*.

It did not take long before Darwin's revolutionary theory was—with some loud exceptions—widely accepted by the scientific community. After all, his theory solved so many mysteries and clarified so much that was previously uncharted or unknown. Of course, Darwin was misunderstood and maligned, but in some cases, he was hated precisely because he was so well understood. Unlike numerous other prominent scientists of his time, Darwin was never knighted by Queen Victoria. Perhaps it was just not possible for the titular head of the Church of England to honor a man some church leaders may have considered a heretic. Interestingly, many mainline religions—willing to view biblical passages somewhat metaphorically—gradually acknowledged evolution by natural selection. This was brought home for me while I was doing genealogy research. I discovered an old letter that my maternal grandfather, a Methodist minister born in 1885, had written to my aunt. In it, he said he not only accepted evolution, but believed it. However, with the rise of religious fundamentalism (now known as the evangelical movement) and its emphasis on biblical inerrancy as well as its insistence on a 4,000 to 6,000-year-old world, Darwin again became a white-hot flash point. He and his theory are maligned and demonized to this day. Darwin's books have never been read because of their literary value, so it would be difficult for me to award him my prize in literature. However, few men have been more consequential than was Charles Darwin. Although unknighted when he died in 1882, Charles Darwin was laid to rest in Westminster Abbey—fittingly near Sir Isaac Newton. Two scientific giants; two theories that changed the world.

The nineteenth century is particularly rich in great authors and great works, but looking only at the 1800s would exclude the author whose name is much more famous now than when he lived, one who is performed, taught, and read as avidly now as ever—William Shakespeare (1564–1616). Around 150 Shakespeare festivals are held annually in the United States alone. It's said that with sales of two to four billion, Agatha Christie may be the top sell-ing author of all time—usually with an asterisk conceding that she may have been outsold only by Shakespeare and the Bible. (Who was keeping track?)

In his book *Draft No. 4: On the Writing Process*, the master writer and teacher John McPhee tells of an experience related to him by the Welsh actor Sir Richard Burton. By the time he was reminiscing with McPhee, Burton—with a voice and looks that other actors could only wish for—had become a famous star of stage and cinema. Burton's story, however, happened back in 1953 when then-twenty-eight-year-old Burton had already done about 60 performances of *Hamlet* at the Old Vic theatre in London. Before he took the stage one evening, the house manager came to Burton's dressing room and advised him: "Be especially good tonight. The old man's out front." Burton inquired, "What old man?" The manager simply said, "He stays for an act and he leaves." A flummoxed Burton again asked, "For God's sake, what old man?" The reply: "Churchill." According to McPhee, as Burton spoke his first line, "he was startled to hear deep identical murmurings from the front row." And that wasn't all. "Churchill continued to follow him line for line, a dramaturgical beetle, his face a thunderhead when something had been cut." Burton remembered, "I tried to shake him off. I went fast. I went slow, he was right there." That night Churchill did not leave after the first act. He stayed "right through to the end." You can see why Burton never forgot that night—he took 18 curtain calls! Over 300 years after his death, the Bard's spirit was still alive at the Old Vic—and in the heart of the man some consider the twentieth-century's greatest statesman.

Starting about 150 years after Shakespeare's death, allegations began to surface that he did not write the works attributed to him—that he was a front man for another writer—and these doubts about Shakespeare are still active today. I suppose some have had difficulty believing that one person could be responsible for the sheer output credited to Shakespeare—along with his acting, producing, etc. Doubters also point to his apparent lack of higher education and to the fact that he did not have the military and other life experiences they imagine would be required to create all those diverse plays. It's also pointed out that there are numerous gaps in Shakespeare's life during which his activities and whereabouts are unknown. (Even his birth year of 1564 is an assumed one.)

The gaps that disturb the doubters are a consolation to others. During those gaps, perhaps he did gain more education; maybe he served in the military—we just don't know. We do know, however, that the citizens of Stratford, by taxing themselves an unusually generous amount, ensured that the grammar school in Stratford was exceptionally good. Students, including Shakespeare, learned

Latin and a smattering of Greek, and they gained a solid appreciation of literature. We also know that Shakespeare used the large, multivolume historical reference *Holinshed's Chronicles* when writing his plays. Was Shakespeare simply such a skilled, creative writer that he could take the history he read about in this comprehensive reference and fashion his dramas—made easier by his occasional lack of concern about historical accuracy?

Three men often promoted as being the "real Shakespeare" are Sir Francis Bacon (1561–1626), Christopher Marlowe (1564–1593), and Edward de Vere, 17th Earl of Oxford (1550–1604). Seven to nine of Shakespeare's plays were written after de Vere's death and twenty-six after Marlowe died. These facts haven't deterred doubters, who propose that the genuine Shakespeare (Marlowe or de Vere) didn't really perish when history records their death. Apparently, instead of dying, they were in hiding—lost to history, families, and friends—writing plays and feeding them to Shakespeare. Bacon, a highly recognized philosopher, statesman, and scientist, outlived Shakespeare by ten years, but was known to be highly egotistical and never shy about taking credit for his many accomplishments. The author James Shapiro says that "it's believed that a third of London's adult population saw a play every month," and Shakespeare's plays were very popular—even with Queen Elizabeth. Failing to take credit for some of the most popular entertainment of their time would have been intolerable for most men (in any era).

It's now difficult to fathom, but at the time of Shakespeare's death, copies of his plays were not in his possession, but scattered among friends, actors, and others. It's unnerving to think how close these plays came to being lost to history. Ben Jonson (1572–1637), another contemporary to whom Shakespeare's works are sometimes attributed, was involved in gathering up the copies into what is now known as *The First Folio*. Jonson, a fellow playwright and Shakespeare's friend, not only made no effort to claim authorship of the plays, but also composed numerous poems in Shakespeare's memory. David Christian in *Origin Story* reminds us that "in Shakespeare's time, even the most educated Europeans generally believed in magic and witchcraft; in werewolves and unicorns; they believed that the Earth stood still . . . that the *Odyssey* was true history." So, considering how long ago they were written, and the audiences they were performed for, how have these plays stood up so well to time? The historian Will Durant helps us understand why Shakespeare's writing overcomes barriers of time and place: "Yes, the plots are improbable, as Tolstoi said; the puns are puerile, the errors of scholarship are un-Baconianly legion, and the philosophy is one of surrender and despair—it does not matter. What matters is that on every page is a godlike energy of soul, and for that we will forgive a man anything."

Of course, the Bard of Avon is not alone in pre-1800s greatness. A few other worthies include Geoffrey Chaucer (c.1340–1400), Miguel de Cervantes (1547–1616), Daniel Defoe (c.1660–1731), Jonathan Swift (1667–1745), and the great Voltaire (1694–1778). Both James Boswell (1740–1795), who wrote *The Life of Samuel Johnson,* and Edward Gibbon (1737–1794), author of *The History of the Decline and Fall of the Roman Empire,* definitely deserve shout-outs; their works are still read, referenced, and discussed in the twenty-first century. *The Wealth of Nations* by Adam Smith (1723–1790), another work from the 1700s, is considered a landmark book that is still studied today. Economics was not even a known academic discipline in Smith's time, and he is sometimes referred to as "The Father of Economics." Rest assured, had I decided to award a prize for economics, Adam Smith would have been a leading candidate. (In 1968, the Nobel Foundation received a donation from Sweden's central bank establishing The Nobel Memorial Prize in Economic Sciences. Since that time, a prize in economics has been administered by the Nobel Foundation and presented with the other Nobel Prizes.)

Two other writers from the 1700s, Henry Fielding (1707–1754) and Laurence Sterne (1713–1768), may have warmed up readers for the great nineteenth-century authors that followed them. When he wrote *Tom Jones*, Fielding entertained his audience while proving that it was acceptable for a character to be bawdy and funny. With *Tristram Shandy*, Sterne brilliantly perfected the fictional memoir; Thomas Jefferson and his wife loved reading Sterne's novel aloud to one another. Speaking of Jefferson, prerevolutionary America and revolutionary America were rich in political writing of extraordinary consequence and eloquence. Just for his prose in the Declaration of Independence, Thomas Jefferson could be considered for the prize. The author Thomas E. Ricks quotes the English writer G. K. Chesterton, who lauded the Declaration as "perhaps the only piece of practical politics that is also theoretical politics and also great literature." Benjamin Franklin's political, ethical, and scientific writings helped secure him a place among the great figures of the Enlightenment, and Thomas Paine's words are still stirring today.

By the nineteenth century in America, however, writing became about more than just politics and freedom. Washington Irving (1783–1859) was one of the first Americans to be recognized by Europeans as an author of note. In addition to the short stories for which he's remembered, Irving was a biographer and historian and served as American ambassador to Spain. It's said that the first American to earn his living primarily as a writer (although not very successfully) was Edgar Allan Poe (1809–1849), who composed short stories, poetry, and literary criticism. Many today remember him for macabre tales like his short story "The Fall of the House of Usher." If you tour the University of Virginia, you can peer into the room where Poe lived when he was a student in Charlottesville. Walt Whitman (1819–1892), with his massive collection of poems, *Leaves of Grass*, is considered one of America's greatest poets. He was not without controversy during his time; some of his poems were judged to be obscene, and he was suspected of being homosexual. His poem about Abraham Lincoln, "Oh Captain! My Captain!" has been recited by countless students and poetry lovers. During the Civil War, Whitman admirably toiled as a nurse in Washington, D.C. hospitals caring for wounded soldiers. Herman Melville (1819–1891) is a now-revered author who felt little of his peers' love or admiration. It's man versus nature in Melville's 1851 novel *Moby Dick*, and for decades some students have felt like it was them against some of Melville's prose. Nevertheless, the book is recognized as one of the finest novels ever written, and it influenced many later authors. The professor and scholar Henry Wadsworth Longfellow (1807–1882) is remembered for poems such as "Song of Hiawatha" and "Paul Revere's Ride." No American poet was more popular than Longfellow during his time.

One center of writing in nineteenth-century-America, and a hotbed of intellectualism, was Concord,

Massachusetts, the home of transcendentalist Ralph Waldo Emerson (1803–1882). An essayist, poet, and philosopher, Emerson was also a renowned speaker who delivered over 1,500 lectures during what must have been a crowded life. Emerson is now also remembered for being the mentor and friend of another Concord author, Henry David Thoreau (Oh, no, not him again!). One of the first classic works that I recall reading was *The Scarlet Letter* by Nathaniel Hawthorne (1804–1864). Yet another New Englander, the shy Hawthorne lived near Emerson and Thoreau for a period of time. Louisa May Alcott (1832–1888) was raised in Concord, where she not only knew Emerson and Hawthorne, but also was tutored by Thoreau! She may have learned more from him than her ABCs, as she would become an abolitionist, a feminist, and an activist in the temperance movement—and like Thoreau, Alcott never married. Alcott's books, starting with *Little* Women, were fictional, but she drew heavily on her experiences growing up in the transcendentalist, intellectual, and sometimes wonderful environment of the Alcott home. Alcott wrote about her imaginary family in a way that has appealed to readers ever since. Hiding out in not-too-distant Amherst, Massachusetts, the reclusive Emily Dickinson (1830–1886), was, for all intents and purposes, unknown as a poet during her lifetime. After her death, Emily's sister found her poems—untitled and written on scraps of paper—in

Emily's bedroom, where she had spent most of her time after college. She is now one of America's most cherished poets. Leaving fiction for a moment, I could not argue against awarding a Nobel Prize to Abraham Lincoln (1809–1865) for his *Gettysburg Address*. Lincoln perished with no awareness that his brief oration would become one of the most admired, quoted, and memorized documents in the American canon.

In nineteenth-century England, novels were ubiquitous, and between 1814 and 1832, the wildly successful Walter Scott (1771–1832) published a novel every 18 months. Besides authoring 20 works of fiction, Scott also distinguished himself as a poet. In 1820, he was made a baronet and has been known since then as "Sir Walter Scott." Forty-one years later, Scott was still popular with the royal family; in 1861, Queen Victoria read Scott's *Peveril of the Peak* to her beloved husband, Prince Albert, as he lay dying in Windsor Castle (she only reached page 81).

Born in Edinburgh, Sir Walter properly belongs to the Scottish Enlightenment. His popular historical works like *Ivanhoe* and *Waverly* had great appeal for male readers and inspired such writers as James Fennimore Cooper and Tolstoy. Beloved as he was, not everyone was or is now a fan. Author Kurt Andersen calls Scott's books overwrought and sentimental. Andersen also claims that American Southerners used Scott's works "to justify and romanticize their slave-based feudalism." He says Southern children, steamboats, and towns were named after characters and imaginary places in Scott's novels. From Anderson, I also learned of Mark Twain's low opinion of Scott. Twain scathingly wrote, "Sir Walter had so large a hand in making Southern character, as it existed before the war, that he is in great measure responsible for the war." Andersen does admit that Twain was prone to hyperbole, and I have to wonder if there was some professional jealousy lurking in Twain's opinion of Scott.

Sir Walter Scott has been popular in lots of venues, but I'll tell you one in which he's not: the men's book club to which I belong. Years ago, it was my turn to select a book for our group to read and discuss. I chose *Ivanhoe* and kiddingly offered a Classic Comic version to anyone who preferred a lighter approach to Scott's novel. There were about 13 members in the club, and on the night we discussed *Ivanhoe*, only five showed up (including me). That attendance remains a 40-plus year record! If my friends are any indication, Scott may not have aged well, but he was one of the most successful writers of his time.

What about those famous sisters, the Brontes? Charlotte (1816–1855) is remembered for *Jane Eyre* and Emily (1818–1848) for *Wuthering Heights*. I learned from Timothy Spurgin in his Great Courses offering, "The English Novel," that *Jane Eyre* insists on the equality of men and women while it protests against the class system." For the non-English major, it's a great love story with some hellishly good writing. Spurgin reminds his students that *Jane Eyre* ended with a kind of poetic justice, whereas Emily's darker *Wuthering Heights* is a true tragedy. Becoming literally acquainted with Mr. Rochester and Heathcliff made this non-English major wonder what sort of men those sisters grew up around! Their father was a Yorkshire minister, but Spurgin says the sisters were heavily influenced by Lord Byron (a bit of a literary "bad-boy" by most accounts), and I will take him at his word. It must not have been easy for women authors in the nineteenth century, and Charlotte may have been speaking for many of them when she lamented, "I wished critics would judge me as an author, not as a woman."

We still read books like *Jane Eyre* and *Wuthering Heights* for many reasons. In his nonfiction work *The Soul of Care*, Arthur Kleinman, a psychiatrist and anthropologist, elegantly writes that "Caregiving is perhaps the most ubiquitous activity of human beings. It is also the existential activity through which we most fully realize our humanity. In the humblest of moments of caring—mopping a sweaty brow, changing a soiled sheet, reassuring an agitated person, kissing the cheek of a love one at the end of life—we may embody the finest versions of ourselves." Beautifully said. In *Jane Eyre*, Charlotte Bronte—no Ph.D., no M.D., and

170 years ago—wrote simply: "There is no happiness like that of being loved by your fellow-creatures, and feeling that your presence is an addition to their comfort."

William Makepeace Thackery (1811–1863), an English author working at the same time as the Bronte sisters, wrote *Vanity Fair*, which Professor Spurgin called "the first great multiplot novel." My non-English curriculum did not include *Vanity Fair*. However, after watching a high-quality dramatization of Thackery's work, I knew that it couldn't have been based on anything but a great epic novel—one that belongs on my list of future reads.

Scott, the Brontes, and Thackery were far from being the only superb nineteenth-century English novelists. Writing a bit earlier than both the Bronte sisters and Thackery was Jane Austen (1775–1817). Her novels, such as *Pride and Prejudice* and *Emma*, did not have the historical sweep of Scott's novels, but Spurgin (in his Great Courses lectures) lauds Austen for "her innovative exploration of human psychology and human consciousness." I suppose this could explain some of the immense popularity her novels enjoy today, but I suspect it's her wonderful dialogue that accounts for much of the modern-day love visited upon Austen. Her books have been the subjects of repeated dramatizations, and writers often use generous portions of Austen's original dialogue in their modern scripts. This is not only because of the charming banter between characters, but also due to the overall elegance and beauty of Austen's writing. It's true, her heroines are lovely and her heroes handsome, but both can also be foolhardy and headstrong—decidedly human. Austen ends her stories with a healthy dose of poetic justice that sees characters receive their just desserts. This means that her characters often achieve not only "forever after" romantic happiness, but also lessons in humility that leave them more sensitive and attuned to the misfortunes and struggles of lesser characters. Austen was one of the first authors to use what is now called interior monologues or stream-of-consciousness narrative, allowing the reader a certain intimacy with her characters. Later authors, such as George Eliot, employed this literary tool even more extensively. Austen's popularity during her lifetime was modest, and according to Spurgin, she was particularly proud of a complimentary notice of *Emma* by Sir Walter Scott. Later writers like George Eliot and Henry James admired Austen and considered her a "writer's writer." Today, Jane is a literary superstar.

We can't neglect the author who gave us some of the most dastardly, some of the most bizarre, and some of the most laudible characters in English literature: Charles Dickens (1812–1870). Non-English majors everywhere are familiar with characters like Oliver Twist, Ebenezer Scrooge, Miss Havisham, and Fagin. It isn't only on the page that these characters come alive; watching skilled actors bring Dickens's bizarre, earthy personalities to the stage or screen can be just as delightful. Dickens had apparently considered acting as a career, which may have helped him shape such colorful characters. However, the nearest that Dickens came to acting was his successful career performing public readings of passages from his books, including two American tours. Dickens has remained a tremendously popular author—at times more beloved in the United States than in Britain. Dickens's 1843 novel, *A Christmas Carol*, revolutionized the Christmas holiday— and the way in which we now celebrate it.

There are readers today who may find Dickens's books— particularly some of the early ones that were serialized—to be overly wordy. Not only was he supposedly paid by the word for these serializations, but Dickens may also have been his own editor. Excess verbiage and all, Dickens's reputation only grows. The critic Robert Gottlieb writes in *The New York Times* that Dickens "as time goes by emerges ever more conclusively as England's greatest novelist and the literary figure who has come to govern our sense of the Victorian era; to embody it, really." Gottlieb's belief is, I'm sure, shared by many.

In his book *The Road to Character*, David Brooks profiled the author George Eliot, among others. Brooks wrote admiringly of Eliot, and I was motivated to try an Eliot novel. I recently looked in my book journal to see what

I had recorded after completing *Silas Marner*, my first venture into Eliot's world. Typically, I write a very brief impression, and this entry was no exception: "Classic— great story and wonderful writing." (High praise from a non-English major!) Next, I tackled *Middlemarch*, the work generally considered to be Eliot's masterpiece. My entry: "Worth every minute it took to read. Huge, epic novel." It appears that I was on the road to becoming a George Eliot devotee.

George Eliot was the career-long pen name of Mary Ann (or Marian) Evans (1819–1880). Eliot's longtime partner was George Henry Lewes, and Eliot sometimes went by "Mrs. Lewes." (Lewes may have been pronounced "Lewis.") George Lewes, a fellow intellectual, was married but unable to secure a divorce from his estranged wife—not rare in those times. Eliot achieved some fame and notoriety in her own lifetime, and their long cohabitation was no secret.

Eliot populated all her novels with memorable characters, but in *Middlemarch*, Eliot outdid herself. Dorothea, the central character, is so captivating, so vulnerable, that you yearn to reach through the pages, grab her by the shoulders, and shout, "Don't do it, don't marry the creep!" Of course, she marries the creep. In his lectures on Eliot, Professor Spurgin claims Dorothea was a new kind of character—complex, with intellectual and spiritual yearnings that "make her something of a misfit." For example, she is rich and beautiful, but dresses down; she has lofty ideals and thoughts, and wants more out of life than others. When I read that Dorothea married a much older man, hoping she would find an intellectual partner to satisfy her yearnings, I was irked. I knew that he was actually a pious, insufferable, intellectual fraud. I think Spurgin would inform me that at least some of my frustration was because Eliot was "not afraid to acknowledge pain, failure, and fear" and that we "can recognize our own weaknesses through her characters." Maybe—or maybe Eliot created such an appealing character that I couldn't bear watching her make this appalling choice.

Eliot is known for her long, authorial interludes where she speaks directly to the reader, but she is also the master of the short descriptor, telling us a lot about someone with just a few words. One of my favorites is from *The Mill on the Floss*: "She wished she could be like Bob, with his easily satisfied ignorance." She could also paint a vivid, rich portrait in just one sentence as she did of Mr. Wrench, a fictitious local doctor in *Middlemarch*. "Mr. Wrench was a small, neat, bilious man, with a well-dressed wig; he had a laborious practice, an irascible temper, a lymphatic wife and seven children."

Eliot, herself, could have been a character in one of her novels. Described by her contemporaries as a very homely woman, she, nevertheless, quickly won over men and women alike—the men often claiming they couldn't help falling in love with her. As a young woman, she read *An Inquiry Concerning the Origin of Christianity* by Charles Hennell and came to view the Bible as "histories consisting of mingled truth and fiction." She continued to admire the teachings of Jesus but could not accept his divinity. Her beliefs caused great family discord when she would no longer accompany her father and brother to church services. Her characters who are members of the clergy can be among her most compelling; this made me wonder how her views on Christianity impacted her writing. Perhaps it allowed her the freedom to portray some of these people of the cloth as quite wonderful souls, but others as reprehensible. Charles Darwin and his family were ardent readers of Eliot's novels, and Darwin visited her on at least one occasion. Was it just polite chit-chat over tea? I hope not. We do know that Eliot read *On the Origin of Species*, and in a letter to a friend wrote that the book would have "great effect in the scientific world, causing a thorough and open discussion of a question about which people hitherto felt timid. So the world gets on step by step toward brave clearness and honesty!"

According to Rebecca Mead in *My Life in Middlemarch*, many consider the final sentence in *Middlemarch* to be "one of the most admired in literature." It reads, "But the effect of her being on those around her was incalculably diffuse:

for the growing good of the world is partly dependent on unhistoric acts; and that things are not so ill with you and me as they might have been, is half owing to the number who lived faithfully a hidden life, and rest in unvisited tombs." You may not agree with the literary critic Stanley Fish who considered the sentence "quietly thrilling," but I suspect readers venture back to nineteenth-century literature for the chance to savor that kind of sentence.

Professor Spurgin taught that in *Middlemarch*, Eliot "perfects the form of the multiplot novel combining the sweep of Dickens and Thackery with the psychological acuity of Austen." He also believes that she "brings an unprecedented intellectual and moral seriousness to the English novel." I read and loved George Eliot's novels before I learned any of the professor's insights—just like generations of other satisfied readers.

Deciding on candidates from which to pick my fanciful literature laureate from the past was easy. Choosing a winner—or winners—was not! My thoughts went back to Oliver Sacks's reminiscence of the girl who didn't have time to read a classic novel: how she might have information, but not knowledge; how she could not really understand human nature and society. Then, I happened to read Helen MacDonald's *Vesper Flights*. In her introduction to these essays, MacDonald writes that "Literature can teach us the qualitative texture of the world. And we need it to. We need to communicate the value of things, so that more of us might fight to save them." Something clicked. Yes. "The qualitative texture of the world"—you can't rely on nonfiction for that. I can see Sacks nodding in agreement.

There are four authors whom I feel were virtuosos when imparting the qualitative texture of the world to their readers, and they accomplished it in ways that were oftentimes delightful, now and again appalling, but always sublime: **William Shakespeare** (1564–1616), **Jane Austen** (1775–1817), **Charles Dickens** (1812–1870), and **Mary Ann Evans** aka **George Eliot** (1819–1880). Recognizing that my prize is a whimsical one—and totally under my auspices—I'm choosing to ignore Nobel Foundation rules (where only three winners can share a prize) and award my Nobel Prize in Literature to each of these four masters of the English language. Congratulations to them all!

Starstuff

In the fifth century BC, the philosopher Democritus
proposed that all matter was made of tiny and
indivisible atoms, which came in various sizes
and textures—some hard and some soft,
some smooth and some thorny.
But the atoms themselves were accepted
as givens, or "first beginnings."
—**Alan Lightman, The Accidental Universe**

*L*ong ago, our ancestors first evolved sufficient intellect to become fully sentient beings, capable of appreciating both the inevitability and the finality of death. Unlike the creatures around them—who were living mostly in the moment—certain of these forebears would have become discomfited by a deep fear of their eventual nonexistence. From this dread of the coming unknown would emerge diverse beliefs and practices that have shaped much of recorded human history.

I know enough about my ancestors to be sure that the afterlife—how to attain it, what form it would take, and whom it would include—was an all-consuming concern for countless numbers of them. After all, disagreeing with powerful religious figures about this topic was for a time in Europe so dangerous that you risked being burned at the stake! Concern with the afterlife, it seems, is not limited to our more current Neolithic era. As far back as Paleolithic times, some early humans were buried with possessions; we presume these tokens were placed there by loved ones to help the deceased make a journey. It's hard to know where these early folks thought that their final journey might take them; some may have anticipated crossing a huge body of water, while others may have seen themselves finally climbing that distant mountain. It would have been natural, I think, for some to gaze upward and imagine the unknown expanse above them—with its huge warming sun and beautiful night sky—as their creator's home, hence their final destination. In spite of Copernicus, Galileo, and telescopes, generations of Christians have clung to this belief. Even nonbelievers still reference the heavens above them.

For almost the entirety of human history, our prede-cessors could not have imagined how mind-numbingly vast the firmament above them really is (or the fact that it is constantly expanding). In *A Universe from Nothing: Why There is Something Rather Than Nothing*, physicist and author Lawrence Krauss tells his readers to venture out to some location at night where the stars are visible and "hold up your hand to the sky, making a tiny circle between your thumb and forefinger about the size of a dime. Hold it up to a dark patch of the sky where there are no visible stars. In that dark patch, with a large enough telescope of the type we now have in service today, you could discern perhaps 100,000 galaxies, each containing a billion stars." This may seem almost impossible until we recognize that our observable universe contains billions of galaxies. Each of these galaxies—like our galaxy, the Milky Way—contains billions of stars. This means that there are more planets than most of us can ever imagine—perhaps 700 million trillion!

Just as astounding as the enormity of the universe is the magnitude of another realm—the quantum world. It is as minute as the universe is expansive. Many of us find it challenging to comprehend a submicroscopic atomic world where single cells contain more than a trillion atoms. Author Richard Dawkins is helpful when he describes it like this in *The Magic of Reality*: ". . . the number of carbon atoms in even the smallest diamond crystal—is gigantic, more than all the fish (plus all the people) in the world." Leonard Mlodinow, author of *The Upright Thinkers*, colorfully writes, "To get an idea how small atoms are, imagine filling all the world's oceans with marbles. And then imagine shrinking each of those marbles down to the size of an atom. How much space would they take up? Less than a teaspoon."

I've been inclined to picture the quantum world as its own universe. This may stem from learning years ago that each atom, with its electrons orbiting around a nucleus, resembled—at least to me—its own tiny solar system. I should abandon this simplistic view of the atom as it was discarded and replaced years ago with a much more complex model. Instead of discrete, unambiguous orbits, this newer model has probabilities and uncertainties. Of course, the quantum world is not its own universe, but rather the very foundation of our larger universe, created over 13 billion years ago in the granddaddy of all quantum nuclear events—the Big Bang.

In his book *Until the End of Time*, physicist Brian Greene tells us: "Grind up anything previously alive, pry apart its molecular machinery, and you'll find an abundance of the same six types of atoms: carbon, hydrogen, oxygen, nitrogen, phosphorus, and sulfur . . ." But it isn't just every living thing that is composed of atoms, it's every single thing—rocks, water, even the air we breathe. One of my favorite books is Alan Lightman's *Searching for Stars on an Island in Maine.* Both a physicist and a philosopher, Lightman reflects, "It is astonishing but true that if I could attach a small tag to each of the atoms of my body and travel with them backward in time, I would find that those atoms originated in particular stars in the sky. Those exact atoms." The inimitable Carl Sagan said it years ago: "We are made of starstuff."

In spite of the astonishing number of atoms there must be in the universe, it ends up that the Cosmos is really quite frugal. The ancient atoms that comprise each of us will live on in future plants, animals, and humans—or in almost anything you might think of. As Bill Bryson wrote in *A Short History of Nearly Everything*: "When we die our atoms will disassemble and move off to find new uses elsewhere—as part of a leaf or other human being or drop of dew." Bryson also says, "We are each so atomically numerous and so vigorously recycled at death that a significant number of our atoms—up to a billion of each of us, it has been suggested—probably once belonged to Shakespeare." Or, I might add, to Jesus Christ, Mohammed, or Charles Darwin!

Compared to the eons of stargazers who have lived before us, we are now able to appreciate so much more about our universe. We recognize that we are all made of "starstuff," and that each one of us is a unique aggregate of multiple short reincarnations. Although we don't feel like it, we are each amazingly 13 billion years old! Understanding these things, I can't help but wonder: Rather than looking upward when searching for immortality, should we instead be looking inward? Perhaps we need look no further than to our very own eternal atoms.

Aging with Grace

I don't feel like an old man, I feel like a young man
with something the matter with him.
(Comment by Bruce Bliven.)
—George Sheehan, **Running and Being**

*L*iving to old age is a gift, but aging is a part of every life, no matter how short. Probably for that reason, sayings about aging are ubiquitous—and often philosophical. A quick trip to Google yields a variety of these aphorisms: "The afternoon knows what the morning never suspects" (Robert Frost); "The wisest are the most annoyed at the loss of time" (Dante); and from George Bernard Shaw: "You don't stop laughing because you grow old, you grow old because you stop laughing." My favorite, by Philip Roth, is more brutal than philosophical: "Old age is not a battle. Old age is a massacre." Ironically, if we do make it to old age, few of us stop to consider that we are the fortunate ones—outliers in most of human history.

Our view of aging depends on our vantage point. Youngsters rarely give much thought to growing old. They're apt to write off anyone much older than they are as impossibly old—even irrelevant. In midlife—years that are often our most productive and vital—we begin to acknowledge the passing years but take refuge in condescending platitudes such as: "You're only as old as you feel." In these robust years, we seem to believe that our refusal to succumb to certain sins or habits will ensure that we never resemble our parents and grandparents. Depending on our particular values and prejudices, the narrative goes something like this: "If I keep hiking," or "If I keep running," or "If I keep golfing," or "If I keep working," then "I'll never look like that!"

If *Homo sapiens* endure long enough, some future Nobel laureate will manipulate DNA telomeres or break the genetic commands for cell death, and aging will be defeated. Years ago, Kurt Vonnegut, Jr., took on this subject when he published *Welcome to the Monkey House*, a humorous collection of short stories. The last story, "Tomorrow and Tomorrow and Tomorrow"—written in 1953—imagines the discovery of an elixir called "anti-gerasone." This inexpensive potion made of mud and dandelions stops aging in its tracks. What follows, naturally, are the to-be-expected issues of overcrowding, strife, resentment, and shortages. The central character, one-hundred-seventy-two-year-old "Gramps,"—who was already seventy when anti-gerasone was discovered—tyrannically rules over his extended family by continually changing his will as he suspects various family members of diluting his anti-gerasone. This leads ninety-three-year-old "Emerald" to lament that, "Sometimes I wish folks just up and died regular as clockwork, without anything to say about it, instead of deciding themselves how long they're going to stay around. There ought to be a law against selling the stuff to anyone over one hundred and fifty."

Regardless of whether they perceive aging as a battle or a massacre, most folks inhabiting the "golden years" will admit that this journey is bound to require giving up certain things. I have noticed that the most adroit agers never relinquish pleasures too easily, but that when they must, they surrender gracefully to the inevitable—and then take care not to make life miserable and guilt-ridden for those left to pick up the pieces. Some things are not difficult to give up with grace. Giving up tennis for pickleball is not gut-wrenching, and walking instead of running should not destroy your spirit. But, what about surrendering the car keys, moving to a more sheltered living space, or adopting a walker?

The clever quip: "Growing old is not for sissies," has always made me smile. I now appreciate what this adage implies: aging may require courage. For some of us, accepting the limitations of age with grace might be our final chance to be that brave soul we always hoped or imagined we could be.

> *My candle burns at both ends;*
> *It will not last the night;*
> *But ah, my foes, and oh, my friends—*
> *It gives a lovely light.*

—Edna St. Vincent Milay, "First Fig"

Finis

Bibliography

A Good Book

Boswell, James. *The Life of Samuel Johnson*. New York: Penguin, 2009.

Durant, Will. *Fallen Leaves: Last Words on Life, Love, War, and God*. New York: Simon & Schuster, 2014.

Durant, Will. *The Greatest Minds and Ideas of All Time*. New York: Simon & Schuster International, 2002.

Johnson, Steven. *Invention of Air: A Story of Science, Faith, Revolution, and the Birth of America*. New York: Riverhead Books, 2009.

The Missing Link

Condemni, Silvana, and Francois Savatier. *A Pocketbook of Human Evolution: How We Became Sapiens*. New York: The Experiment, 2019.

Foundation, The Leakey. "Origin Stories Podcast. 31 Oct. 2019, Episode 39, podcasts.apple.com/us/podcast/origin-stories.

Gurche, John, et al. *Lost Anatomies: The Evolution of the Human Form*. New York: Abrams, 2019.

Johanson, Donald, and Kate Wong. *Lucy's Legacy*. New York: Harmony Books, 2009.

Pattison, Kermit. *Fossil Men: The Quest for the Oldest Skeleton and the Origins of Humankind*. New York: HarperCollins, 2020

Shubin, Neil. *Some Assembly Required: Decoding Four Billion Years of Life, from Ancient Fossils to DNA*. New York: Pantheon, 2020.

Shubin, Neil. *Your Inner Fish: A Journey into the 3.5-Billion-Year History of the Human Body*. New York: Pantheon, 2008.

My Ancestors

—. "Abigail Adams Smith." History of American Women Blog, Retrieved August 17, 2020.

Childs, Craig, and Sarah Gilman. *Atlas of a Lost World: Travels in Ice Age America*. New York: Pantheon, 2018.

Christian, David. *Maps of Time: An Introduction to Big History*. Berkeley: University of California Press, 2004.

Condemni, Silvana, and Francois Savatier. *A Pocketbook of Human Evolution: How We Became Sapiens*. New York: The Experiment, 2019.

Cozzens, Peter. *The Earth Is Weeping: The Epic Story of the Indian Wars for the American West*. New York: Vintage, 2016.

Cross, John David. *The First Americans*. Novato: New World City Publishing, 2016.

Eaton, S. Boyd, and Melvin Konner. "Paleolithic Nutrition: A Consideration of Its Nature and Current Implications." *New England Journal of Medicine*, vol. 312, no. 3, 31 Jan. 1985, pp. 283–289.

Foundation, The Leakey. "Origin Stories Podcast."10 Nov. 2020, Episode 47, podcasts.apple.com/us/podcast/origin-stories.

Goldstone, Lawrence, and Nancy Bazelon Goldstone. *Out of the Flames: The Remarkable Story of a Fearless Scholar, a Fatal Heresy, and One of the Rarest Books in the World*. New York: Broadway Books, 2003.

Gurche, John, et al. *Lost Anatomies: The Evolution of the Human Form*. New York: Abrams, 2019.

Harari, Yuval N. *Sapiens: A Brief History of Humankind*. New York: Harper, 2015.

Hawks, John. "The Rise of Humans: Great Scientific Debates." The Great Courses.

McClellan, James E. *Science and Technology in World History*. Baltimore: Johns Hopkins University Press, 2006.

McCullough, David. *John Adams*. New York: Simon & Schuster, 2001.

Nerburn, Kent. *Chief Joseph & the Flight of the Nez Perce: The Untold Story of an American Tragedy*. San Francisco: HarperOne, 2006.

Robinson, Jo. *Eating on the Wild Side: The Missing Link to Optimum Health*. New York: Little Brown & Co, 2013.

Shubin, Neil. *Some Assembly Required: Decoding Four Billion Years of Life from Ancient Fossils to DNA*. New York: Pantheon, 2020.

Wells, Laura Dassow. *Henry David Thoreau: A Life*. Chicago: The University of Chicago Press, 2018.

Wilson, Edward O., and Debby Cotter Kaspari. *Genesis: The Deep Origin of Societies*. New York: Liveright Publishing Corporation, 2019.

Wrangham, Richard W. *Catching Fire: How Cooking Made Us Human*. New York: Basic Books, 2009.

TRUSTING HISTORY

Caro, Robert A. *Working*. New York: Knopf, 2019.

McCullough, David. *Pioneers: The Heroic Story of the Settlers Who Brought the American Ideal West*. New York: Simon & Schuster, 2020.

Skidmore, Chris. *Richard III: England's Most Controversial King*. New York: St. Martin's Press, 2018.

Tey, Josephine. *The Daughter of Time*. New York: Simon & Schuster, 1995.

MARCHING THROUGH HISTORY

Barnosky, Anthony D. *Dodging Extinction - Power, Food, Money, and the Future of Life on Earth*. Berkeley: University of California Press, 2014.

Heine, Steven J. *DNA is Not Destiny: The Remarkable, Completely Misunderstood Relationship between You and Your Genes*. New York: W.W. Norton & Company, 2017.

Kenneally, Christine. *The Invisible History of the Human Race: How DNA and History Shape Our Identities and Our Futures*. London: Penguin Books, 2015.

Pont, Donald E. *A Family Journey*. Phoenix: 1106 Design, 2012

Rutherford, Adam. *A Brief History of Everyone Who Ever Lived: The Human Story Retold Through Our Genes*. London: Weidenfield & Nicolson, 2017.

Sykes, Bryan. *DNA USA: A Genetic Portrait of America*. New York: Liveright Publishing Corp, 2013

AMERICAN DREAM

Pont, Donald E. *A Family Journey*. Phoenix: 1106 Design, 2012.

AMERICAN EXCEPTIONALISM

Drury, Bob, and Tom Clavin. *Valley Forge*. New York: Simon & Schuster, 2019.

Jenkinson, Clay. *Repairing Jefferson's America*. Virginia Beach: Koehler Books, 2020.

Kors, Alan Charles. "Voltaire and the Triumph of the Enlightenment." The Great Courses.

McCullough, David. *John Adams*. New York: Simon & Schuster, 2001.

Paine, Thomas. *Common Sense*. Amazon.com Services, 2012.

Paine, Thomas. *The Age of Reason*. Rainbow Classics, 2016.

Paine, Thomas. *The American Crisis*. Amazon.com Services, 2020.

Robinson, Daniel N. "American Ideals: Founding a Republic of Virtues." The Great Courses.

"A World of Paine." *Revolutionary Founders: Rebels, Radicals, and Reformers in the Making of the Nation*, by Alfred Fabian Young and Jill B. Lepore, New York: Knopf, 2011.

THE GREATEST AMERICAN

Adams, Abigail. "Letter from Abigail Adams to John Adams." 16 July 1775.

Adams Family Papers: Massachusetts Historical Society: An Electronic Archive

Chaffin, Tom. *Revolutionary Brothers: Thomas Jefferson, the Marquis de Lafayette, and the Friendship That Helped Forge Two Nations*. New York: St. Martin's Press, 2019.

Ellis, Joseph J. *Revolutionary Summer: The Birth of American Independence*. Vintage Books, 2014.

Ellis, Joseph J. *The Quartet: Orchestrating the Second American Revolution, 1783-1789*. New York: Random House, 2016.

McCullough, David. *1776*. New York: Simon & Schuster, 2005.

Philbrick, Nathaniel. *Valiant Ambition George Washington, Benedict Arnold, and the Fate of the American Revolution*. New York: Viking, 2016.

Vowell, Sarah. *Lafayette in the Somewhat United States*. Riverhead Books, 2016.

FOR KATIE

Chernow, Ron. *Alexander Hamilton*. London: Penguin Books, 2005.

WISDOM OF THE FOUNDERS

Adams, John, et al. *My Dearest Friend: Letters of Abigail and John Adams*. Cambridge: Belknap Press of Harvard University Press, 2010.

Boles, John B. *Jefferson - Architect of American Liberty*. New York: Basic Books, 2017.

Hitchens, Christopher. *Thomas Jefferson: Author of America*. HarperCollins e-Books, 2009.

Jenkinson, Clay. *Repairing Jefferson's America*. Virginia Beach: Koehler Books, 2020.

McCullough, David G. *The American Spirit: Who We Are and What We Stand For*. New York: Simon & Schuster, 2017.

McCullough, David. *John Adams*. New York: Simon & Schuster, 2001.

Meacham, Jon. *Thomas Jefferson: The Art of Power*. New York: Random House, 2012.

Wood, Gordon S. *Friends Divided: John Adams and Thomas Jefferson*. London: Penguin Press, 2017.

The Sins of Our Fathers

Cozzens, Peter. *The Earth is Weeping: The Epic Story of the Indian Wars for the American West*. New York: Vintage, 2016.

Hawks, John. "The Rise of Humans: Great Scientific Debates." The Great Courses.

Horn, Jonathan. *The Man Who Would Not Be Washington: Robert E. Lee's Civil War and His Decision That Changed American History*. Scribner, 2016.

Nerburn, Kent. *Chief Joseph & the Flight of the Nez Perce: The Untold Story of an American Tragedy*. San Francisco: HarperOne, 2009.

The Laureates

Eagleman, David. *Sum: Forty Tales from the Afterlives*. New York: Vintage, 2010.

Peace

Barry, John M. *Roger Williams and the Creation of the American Soul: Church, State, and the Birth of Liberty*. London: Penguin Books, 2012.

Cramer, Jeffrey S. *Solid Seasons: The Friendship of Henry David Thoreau and Ralph Waldo Emerson*. Berkeley: Counterpoint, 2019.

Morris, Edmund. *The Rise of Theodore Roosevelt*. New York: Random House, 2001.

Purkayastha, Shorbori. "Why Mahatma Gandhi was Never Awarded the Nobel Peace Prize." *The Quint*, Retrieved August 20, 2020, www.thequint.com/.

Thoreau, Henry David. *Walden*. Public Domain.

Thoreau, Henry David. *Walking*. Amazon.com Services, 2020.

Walls, Laura Dassow. *Henry David Thoreau: A Life*. Chicago: University of Chicago Press, 2018

Physics

Burke, James. *The Day the Universe Changed*. New York: Little, Brown and Company, 1995.

Christian, David. *Maps of Time an Introduction to Big History*. Berkeley: University of California Press, 2004.

Christian, David. *Origin Story: A Big History of Everything*. New York: Penguin Books, 2019.

Livio, Mario. *Galileo and the Science Deniers*. New York: New York: Simon & Schuster, 2020.

Chemistry

Bensade-Vincent, Benadete. "Dmitri Mendeleev." *Encyclopedia Brittanica*, Febr. 4, 2020. Retrieved August 20, 2020

Burke, James. *The Day the Universe Changed*. New York: Little, Brown & Company, 1995.

Johnson, Steven. *Invention of Air: A Story of Science, Faith, Revolution, and the Birth of America*. New York: Riverhead Books, 2009.

Principe, Lawrence. "Robert Boyle: Anglo-Irish Philosopher and Writer." *Encyclopedia Brittanica*, March 2, 2020. Retrieved August 18, 2020

Schneider, David. *The Invention of Surgery: A History of Modern Medicine from the Renaissance to the Implant Revolution*. New York: Pegasus Books, 2020.

Medicine/Physiology

Ackerknecht, Edward. *A Short History of Medicine*. New York: The Ronald Press Company, 1968.

Barry, John M. *The Great Influenza: The Story of the Deadliest Pandemic in History*. Viking Penguin Books, 2004.

Fitzharris, Lindsey. *The Butchering Art: Joseph Lister's Quest to Transform the Grisly World of Victorian Medicine*. London: Penguin Books, 2018.

Jefferson, Thomas. Letter to Dr. Edward Jenner. 14 May 1806.

Johnson, Victoria. *American Eden: David Hosack, Botany, and Medicine in the Garden of the Early Republic*. Liveright Publishing Corporation, 2019.

Kobler, John. *The Reluctant Surgeon: A Biography of John Hunter*. New York: Doubleday, 1960.

Mann, Charles C. *1491: New Revelations of the Americas before Columbus*. New York: Knopf, 2005.

Osler, Sir William. *The Evolution of Modern Medicine*. New Haven: Yale University Press, 2002.

Pattison, Kermit. *Fossil Men: The Quest for the Oldest Skeleton and the Origins of Humankind*. New York: HarperCollins, 2020

Philbrick, Nathaniel. *Mayflower*. New York: Penguin Books, 2007.

Riedel, Stefan. "Edward Jenner and the History of Smallpox and Vaccination." *Baylor University Medical Center Proceedings*, Jan. 2005.

Schneider, David. *Invention of Surgery*. New York: Pegasus Books, 2020.

LITERATURE

Andersen, Kurt. *Fantasyland: How America Went Haywire:a 500-Year History*. New York: Random House, 2017.

Bakewell, Sarah. *How to Live: Or a Life of Montaigne in One Question and Twenty Attempts at an Answer*. New York: Other Press, 2011.

Bronte, Charlotte. *Jane Eyre*. Beelzebub Classics.

Brooks, David. *The Road to Character*. New York: Random House, 2016.

Christian, David. *Origin Story: A Big History of Everything*. New York: Penguin Books, 2019.

Cramer, Jeffrey. *Solid Seasons: The Friendship of Henry David Thoreau and Ralph Waldo Emerson*. Berkeley: Counterpoint, 2019.

Darwin Correspondence Project, University of Cambridge, www.darwinproject.ac.uk/.

Durant, Will. *The Greatest Minds and Ideas of All Time*. New York: Simon & Schuster, 2002.

Garner, Dwight. *Garner's Quotations: A Modern Miscellany*. New York: Farrar, Straus, and Giroux, 2020

Gottlieb, Robert. Review of *The Mystery of Charles Dickens* by A. N. Wilson. *The New York Times*, 6 November 2020.

Grant, Ulysses S. *Personal Memoirs of U.S. Grant*. RareBooksClub.Com, 2012.

Karp, Walter. *Charles Darwin*. New Word City, 2016.

Kleinman, Arthur. *Soul of Care: The Moral Education of a Doctor*. New York: Viking Press, 2019.

McPhee, John. *Draft No. 4: On the Writing Process*. New York: Straus and Giroux, 2017.

Mead, Rebecca. *My Life in Middlemarch*. New York: Broadway Books, 2014.

Orlean, Susan. *The Library Book*. New York: Simon and Schuster, 2018.

Pont, Donald E. *A Family Journey*. Phoenix: 1106 Design, 2012

"Prince Albert: A Victorian Hero Revealed." PBS, June 14, 2020

Ricks, Thomas E. *First Principles: What American Founders Learned from the Greeks and Romans and How That Shaped Our Country.* New York: HarperCollins, 2020

Sacks, Oliver. *Everything in Its Place: First Loves and Last Tales.* London: Picador , 2020.

Sacks, Oliver. *River of Consciousness.* London: Picador, 2018.

Shapiro, James. *A Year in the Life of William Shakespeare: 1599.* New York: HarperCollins Publishers, 2006.

Sorensen, Ken. "William Shakespeare and His Competition." Osher Lifelong Learning Institute, 24 Sept. 2020, Arizona State University

Thoreau, Henry David. *Walden.* Public Domain

Thoreau, Henry David. *Walking.* Amazon.com Services, 2020.

Walls, Laura Dassow. *Henry David Thoreau: A Life.* Chicago: University of Chicago Press, 2018.

STARSTUFF

Bryson, Bill. *A Short History of Nearly Everything.* New York: Broadway Books, 2003.

Christian, David. *Maps of Time: An Introduction to Big History.* Berkeley: University of California Press, 2004.

Dawkins, Richard. *The Magic of Reality: How We Know What's Really True.* New York: Free Press, 2012.

Krauss, Lawrence Maxwell. *A Universe from Nothing: Why There is Something Rather Than Nothing.* New York: Atria Books, 2012.

Lightman, Alan P. *Searching for Stars on an Island in Maine.* New York: Pantheon Books, 2018.

Miller, Kenneth R. *The Human Instinct: How We Evolved to Have Reason, Consciousness, and Free Will.* New York: Simon & Schuster, 2018.

Mlodinow, Leonard. *The Upright Thinkers: The Human Journey from Living in Trees to Understanding the Cosmos.* New York: Vintage, 2015.

AGING WITH GRACE

Vonnegut, Kurt. *Welcome to the Monkey House.* New York: Dell, 1968.

About the Author

$\mathcal{D}$onald E. Pont was born in Fairbury, Nebraska, in 1944. He is a graduate of the University of Nebraska in Lincoln and the University of Nebraska College of Medicine in Omaha. After serving as a flight medical officer in the U.S. Air Force, he completed a residency in Family Medicine at the University of Minnesota in Minneapolis. Don is a retired family medicine and hospice physician who resides in Mesa, Arizona, with his wife, Jacquelyn.

Made in the USA
Thornton, CO
10/05/22 16:49:20

cceb3ba6-3f1e-44c2-8a74-d66202ef6f53R01